Mechanics
of Functional
Materials

Mechanics
of Functional
Materials

Jiashi Yang
University of Nebraska-Lincoln, USA

World Scientific

NEW JERSEY · LONDON · SINGAPORE · BEIJING · SHANGHAI · HONG KONG · TAIPEI · CHENNAI · TOKYO

Published by

World Scientific Publishing Co. Pte. Ltd.

5 Toh Tuck Link, Singapore 596224

USA office: 27 Warren Street, Suite 401-402, Hackensack, NJ 07601

UK office: 57 Shelton Street, Covent Garden, London WC2H 9HE

Library of Congress Cataloging-in-Publication Data

Names: Yang, Jiashi, 1956- author.
Title: Mechanics of functional materials / Jiashi Yang, University of
  Nebraska-Lincoln, USA.
Description: New Jersey : World Scientific, [2023] | Includes
  bibliographical references and index.
Identifiers: LCCN 2022040023 | ISBN 9789811266010 (hardcover) |
  ISBN 9789811266027 (ebook) | ISBN 9789811266034 (ebook other)
Subjects: LCSH: Deformations (Mechanics) | Materials--Electric properties.
  | Materials--Thermal properties.
Classification: LCC TA417.6 .Y34 2023 | DDC 620.1/123--dc23/eng/20221020
LC record available at https://lccn.loc.gov/2022040023

British Library Cataloguing-in-Publication Data
A catalogue record for this book is available from the British Library.

For any available supplementary material, please visit
https://www.worldscientific.com/worldscibooks/10.1142/13133#t=suppl

Desk Editors: Sanjay Varadharajan/Amanda Yun

Typeset by Stallion Press
Email: enquiries@stallionpress.com

Preface

Conventional mechanics of materials books treat elastic deformations of solids through one-dimensional models for the extension of rods, torsion of shafts and bending of beams. In functional materials, mechanical, thermal, electric and magnetic fields interact. This book presents a systematic treatment of the three-dimensional theories for these coupled phenomena and the corresponding one-dimensional models for extension of rods, torsion of shafts and bending of beams with multi-physical couplings. The scope is limited to linear couplings. The organization of the book is unique. It is extensive, self-contained and concise.

Chapters 1–4 are on the development of three-dimensional theories of elastic, thermal, electric and magnetic fields as well as their interactions. Three-dimensional theories are necessary for a complete understanding of these interactions. For applications in devices such as transducers and sensors, numerical methods based on three-dimensional theories are used very often. Specifically, Chapter 1 is on mechanical behaviors only with a brief summary of conventional one-dimensional mechanics of materials followed by the three-dimensional theory of elasticity. Chapter 2 presents heat conduction in rigid solids and the coupled theory of thermoelasticity. Similar to Chapter 1, the presentation of Chapter 2 also begins with one-dimensional models and then generalizes them to three-dimensional theories. Chapter 3 begins with electrostatics and magnetostatics, followed by coupled theories of magnetoelectricity, pyroelectricity, pyromagnetism and electrodynamics. Dielectrics, conductors and semiconductors are all discussed. Chapter 4 is on more complicated interactions involving mechanical, thermal, electric and magnetic fields.

Chapters 5–7 present one-dimensional models for extension of rods, torsion of shafts and bending of beams systematically with various couplings. In applications, theoretical analyses using one-dimensional models are often possible. One-dimensional models can be constructed directly as in conventional mechanics of materials. On the other hand, they can also be systematically derived from three-dimensional theories. A mixed approach is adopted in this book. The one-dimensional field equations such as the linear or angular momentum equations and the charge equation of electrostatics are established directly from one-dimensional models in most cases. The three-dimensional strain-displacement relations are used to calculate some relevant strain components of the one-dimensional models. The one-dimensional constitutive relations are always reduced from the three-dimensional ones.

The literature on functional materials is numerous. Only those books or papers whose results are directly used in the present book are listed as references. No attempt was made to review the literature. Because of the multi-physical fields involved, for convenience a list of symbols is given in Appendix 1. Some common material constants are gathered in Appendix 2.

Jiashi Yang
July, 2022

About the Author

Jiashi Yang is a Full Professor at the Department of Mechanical and Materials Engineering at the University of Nebraska–Lincoln. He received his bachelor's and master's degrees from Tsinghua University in 1982 and 1985 respectively, and his Ph.D. from Princeton University in 1994. His research interests include the mechanics of electromechanical structures and devices. He served as an Associate Editor for the IEEE Transactions on Ultrasonics, Ferroelectrics, and Frequency Control during 2004–2020. His previous books include *An Introduction to the Theory of Piezoelectricity* with Springer, *Analysis of Piezoelectric Semiconductor Structures* with Springer, and *Mechanics of Piezoelectric Structures* with World Scientific.

Contents

Chapter 1

Mechanics of Materials and Elasticity

This chapter begins with a concise review of the usual mechanics of materials [1] including the direct constructions of one-dimensional theories for extension of rods, torsion of circular shafts and the Euler–Bernoulli theory for bending of beams without shear deformation. Then the two-dimensional theory of plane-stress elasticity is established as a transition to the three-dimensional theory of elasticity that follows. The Cartesian tensor notation and the matrix notation are introduced for the three-dimensional theory of elasticity so that the theory can be used conveniently in anisotropic materials and generalized in later chapters to include thermal, electric and magnetic couplings. The Timoshenko theory for bending of beams with shear deformation is presented in the last section as an example of a general procedure for deriving one-dimensional beam theories from the three-dimensional theory of elasticity.

1.1 Extension of Rods

Consider the slender rod of an isotropic material in Fig. 1.1. The axial coordinate is x. The length of the rod is much larger than the characteristic dimension of its cross-section. The shape of the cross-section is arbitrary. The mass density is ρ. The cross-sectional area is A.

Consider the differential element of the rod in Fig. 1.2. To describe the extension of the rod, we introduce a one-dimensional axial displacement field $u_x = u(x,t)$ which is uniform over a

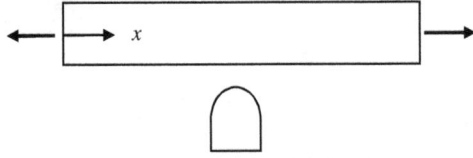

Fig. 1.1.   A rod in extension and its cross-section.

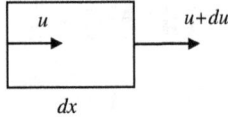

Fig. 1.2.   A differential element of the rod.

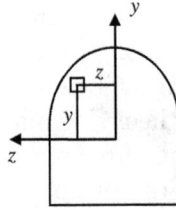

Fig. 1.3.   Centroidal and principal coordinate system in a cross-section.

cross-section [2,3]. The displacements at the two ends of the element are u and $u + du$, respectively.

The axial strain of the rod can be calculated from u through

$$\varepsilon_x = \varepsilon = \frac{(u + du) - u}{dx} = \frac{\partial u}{\partial x} = u', \qquad (1.1.1)$$

where a prime represents a partial derivative with respect to x. Using the stress–strain relation for uniaxial stress (Hooke's law), the axial stress is given by

$$\sigma_x = \sigma = E\varepsilon = Eu', \qquad (1.1.2)$$

which is uniform over a cross-section. E is Young's modulus. Within a cross-section, with respect to the centroidal coordinate system in Fig. 1.3, the only nonzero resultant of the stress distribution in

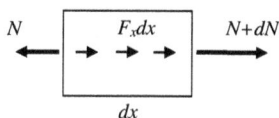

Fig. 1.4. A differential element of the rod under axial loads.

Eq. (1.1.2) is an axial force N with the following expression:

$$N = \int_A \sigma \, dA = \int_A Eu' \, dA = Eu' \int_A dA = EAu'. \qquad (1.1.3)$$

By applying Newton's second law to the free-body diagram of the rod element in Fig. 1.4 in the axial direction, we obtain

$$(N + dN) - N + F_x \, dx = \rho A \, dx \frac{\partial^2 u}{\partial t^2}, \qquad (1.1.4)$$

or

$$\frac{\partial N}{\partial x} + F_x = \rho A \ddot{u}, \qquad (1.1.5)$$

where $F_x(x, t)$ is the axial load per unit length of the rod. A superimposed dot denotes the partial derivative with respect to time. Substituting Eq. (1.1.3) into Eq. (1.1.5), we arrive at a single equation for u, as follows:

$$(EAu')' + F_x = \rho A \ddot{u}. \qquad (1.1.6)$$

For a homogeneous rod with a constant EA, Eq. (1.1.6) reduces to

$$EAu'' + F_x = \rho A \ddot{u}. \qquad (1.1.7)$$

As an example of the application of Eq. (1.1.7), consider waves propagating in an unbounded free rod ($F_x = 0$) described by

$$u = U \exp[i(kx - \omega t)], \qquad (1.1.8)$$

where U is the wave amplitude, i the imaginary unit, k the wave number and ω the wave frequency. The substitution of Eq. (1.1.8) into Eq. (1.1.7) yields

$$-EAUk^2 = -\rho AU\omega^2, \qquad (1.1.9)$$

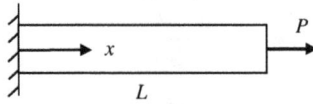

Fig. 1.5.   Extension of a finite rod by an end force.

which determines the speed of extensional waves in the rod as

$$c = \frac{\omega}{k} = \sqrt{\frac{E}{\rho}}. \tag{1.1.10}$$

c is independent of ω or k. Such a wave is said to be nondispersive.

As another example, consider the static extension of a finite rod by an end force P (see Fig. 1.5). The left end is fixed. The boundary-value problem for u is

$$EAu'' = 0, \quad 0 < x < L,$$
$$u(0) = 0, \quad N(L) = P. \tag{1.1.11}$$

Its solution is

$$u = \frac{P}{EA}x, \quad N = P. \tag{1.1.12}$$

Finally, we note that from the three-dimensional point of view the extension of thin rods are characterized by that the lateral stresses are approximately zero, i.e.,

$$\sigma_y = \sigma_z \cong 0. \tag{1.1.13}$$

However, the lateral strains are not negligible (Poisson's effect) and are related to the axial strain by

$$\varepsilon_y = \varepsilon_z = -\nu\varepsilon_x, \tag{1.1.14}$$

where ν is Poisson's ratio. Equation (1.1.14) can be used to determine the lateral displacements u_y and u_z.

1.2   Torsion of Circular Shafts

Consider torsion of the circular shaft of radius R in Fig. 1.6. The twisting moment (torque) is represented by a vector according to the right-hand rule.

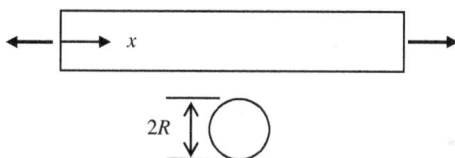

Fig. 1.6. A circular shaft in torsion and its cross-section.

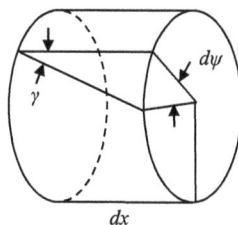

Fig. 1.7. Shear strain γ at a distance r from the axis.

In the torsion of a circular shaft, a cross-section at x rotates around its center as described by an angle of twist function $\psi(x,t)$. When the two cross-sections on the left and right of the differential element of the shaft in Fig. 1.7 with radius $r < R$ have a relative rotation $d\psi$, a shear strain γ is produced which is related to ψ through [1]

$$\gamma\, dx = r\, d\psi, \tag{1.2.1}$$

or

$$\gamma = r\frac{d\psi}{dx}. \tag{1.2.2}$$

Then the shear stress τ over a cross-section is given by

$$\tau = G\gamma = Gr\frac{d\psi}{dx}, \tag{1.2.3}$$

where G is the shear modulus. Within a cross-section, using polar coordinates, the torque M can be calculated from τ according to

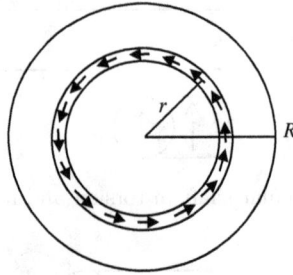

Fig. 1.8. Calculation of M from the shear stress over a cross-section.

Fig. 1.8 as

$$M = \int_0^R Gr\frac{d\psi}{dx}(2\pi r dr)r$$

$$= G\frac{d\psi}{dx}\int_0^R 2\pi r^3 dr = G\frac{d\psi}{dx}I_p = GI_p\psi', \tag{1.2.4}$$

where I_p is the polar moment of inertia of the cross-section about its center:

$$I_p = \int_0^R r^2(2\pi r dr) = \frac{\pi}{2}R^4. \tag{1.2.5}$$

The above analysis is also valid for a circular tube with inner radius a and outer radius b for which

$$I_p = \frac{\pi}{2}(b^4 - a^4). \tag{1.2.6}$$

The moment equation of the differential element in Fig. 1.9 leads to [2,3]

$$(M + dM) - M + m_x\, dx = \rho\, dx I_p\frac{\partial^2\psi}{\partial t^2}, \tag{1.2.7}$$

or

$$\frac{dM}{dx} + m_x = \rho I_p\frac{\partial^2\psi}{\partial t^2}, \tag{1.2.8}$$

where $m_x(x,t)$ represents distributed torque per unit length of the shaft.

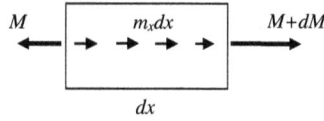

Fig. 1.9.   A differential element under mechanical loads.

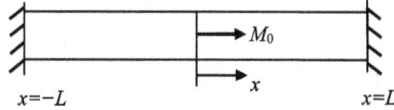

Fig. 1.10.   A shaft under a concentrated torque in the middle.

Substituting from Eq. (1.2.4), we write Eq. (1.2.8) as an equation for ψ:

$$(GI_p\psi')' + m_x = \rho I_p \ddot{\psi}. \tag{1.2.9}$$

In the case of a homogeneous shaft with a constant GI_p, Eq. (1.2.9) reduces to

$$GI_p\psi'' + m_x = \rho I_p \ddot{\psi}. \tag{1.2.10}$$

As an example, consider waves propagating in a free and unbounded shaft described by

$$\psi = \Psi \exp[i(kx - \omega t)]. \tag{1.2.11}$$

The substitution of Eq. (1.2.11) into Eq. (1.2.10) yields

$$-GI_p\Psi k^2 = -\rho I_p \Psi \omega^2, \tag{1.2.12}$$

which determines the torsional wave speed as

$$\frac{\omega}{k} = \sqrt{\frac{G}{\rho}}. \tag{1.2.13}$$

As another example, consider the torsion of a shaft under a concentrated torque M_0 in the middle as shown in Fig. 1.10. The two ends are fixed.

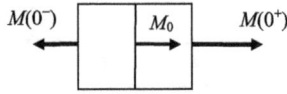

Fig. 1.11.   A differential element in the middle of the shaft.

Because of the presence of a concentrated torque, the basic approach is to analyze the two halves of the shaft separately and then apply boundary conditions at the two ends and continuity or jump conditions in the middle. The boundary-value problem is

$$
\begin{aligned}
&GI_p\psi'' = 0, \quad -L < x < 0, \\
&GI_p\psi'' = 0, \quad 0 < x < L, \\
&\psi(-L) = 0, \quad \psi(L) = 0, \\
&\psi(0^-) = \psi(0^+), \quad M(0^+) + M_0 - M(0^-) = 0,
\end{aligned}
\tag{1.2.14}
$$

where the jump condition (the last one in Eq. (1.2.14)) is based on the moment equation of the differential element taken in the middle of the shaft as shown in Fig. 1.11.

The solution of the first two equations of Eq. (1.2.14) is

$$
\begin{aligned}
\psi = C_1(x + L), \quad M = GI_pC_1, \quad -L < x < 0, \\
\psi = C_2(x - L), \quad M = GI_pC_2, \quad 0 < x < L,
\end{aligned}
\tag{1.2.15}
$$

which satisfies the boundary conditions at the two ends. C_1 and C_2 are determined by the continuity and jump conditions in the middle of the shaft:

$$
\begin{aligned}
&C_1L = C_2(-L), \\
&GI_pC_2 + M_0 - GI_pC_1 = 0,
\end{aligned}
\tag{1.2.16}
$$

or

$$
C_1 = -C_2 = \frac{M_0}{2GI_p}.
\tag{1.2.17}
$$

1.3   Euler–Bernoulli Theory for Bending of Beams

Consider the beam in Fig. 1.12. For bending in the (x, y) plane, we assume that the cross-section is symmetric about the y-axis. When a

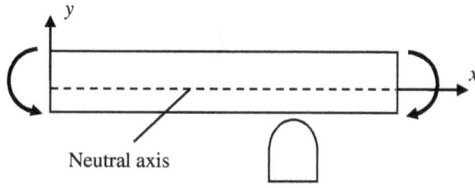

Fig. 1.12.   A beam in bending and its cross-section.

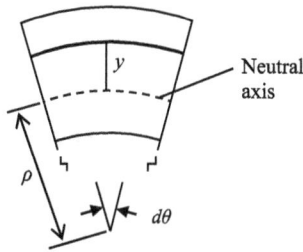

Fig. 1.13.   A differential element for bending deformation.

beam is in bending, the upper part of the beam is in axial extension and the lower part in compression or vice versa. There exists a plane that is neither stretched nor compressed (neutral plane) where the x-axis is placed (neutral axis). Its location is to be determined as follows.

From the differential element of the beam in Fig. 1.13, the extensional strain of an axial fiber of the beam at coordinate y from the neutral axis can be calculated as

$$\varepsilon = \frac{(\rho + y)d\theta - \rho\,d\theta}{\rho\,d\theta} = \frac{y}{\rho}, \tag{1.3.1}$$

where ρ is the radius of curvature of the neutral axis.

Then the axial stress is given by

$$\sigma = E\varepsilon = E\frac{y}{\rho}. \tag{1.3.2}$$

For bending, the axial force over a cross-section is

$$N = \int_A \sigma\,dA = \int_A E\frac{y}{\rho}dA = \frac{E}{\rho}\int_A y\,dA = 0, \tag{1.3.3}$$

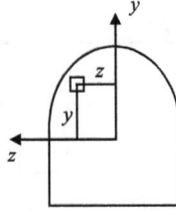

Fig. 1.14.   Centroidal and principal coordinate system in a cross-section.

which determines that the neutral axis is a centroidal axis. The bending moment over a cross-section can be calculated as (see Fig. 1.14)

$$M = \int_A y\sigma \, dA = \frac{E}{\rho} \int_A y^2 \, dA = \frac{EI}{\rho}, \qquad (1.3.4)$$

where I is the moment of inertia of the cross-section about the z-axis:

$$I = \int_A y^2 \, dA. \qquad (1.3.5)$$

Let the displacement function (deflection curve) of the neutral axis be $v = v(x, t)$ which is assumed to be small, the radius of curvature can be written approximately as

$$\frac{1}{\rho} = \frac{\pm v''}{[1 + (v')^2]^{3/2}} \cong -v'', \qquad (1.3.6)$$

where the minus sign has been chosen for consistency with the definition of the bending moment in Eq. (1.3.4). From Eqs. (1.3.4) and (1.3.6),

$$M = -EIv''. \qquad (1.3.7)$$

From the equation of motion in the y direction of the differential element in Fig. 1.15 and its moment equation about, e.g., the center of its right face, we obtain

$$-Q + F_y \, dx + (Q + dQ) = \rho A \, dx \frac{\partial^2 v}{\partial t^2},$$
$$M + Q \, dx - (M + dM) \cong 0, \qquad (1.3.8)$$

where the rotatory inertia of the element has been neglected in the moment equation.

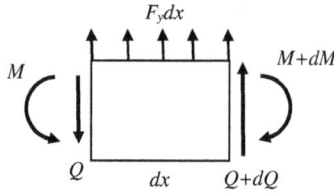

Fig. 1.15.   A differential element with mechanical loads for bending.

Equation (1.3.8) can be rewritten as

$$\frac{\partial Q}{\partial x} + F_y = \rho A \frac{\partial^2 v}{\partial t^2},$$

$$\frac{\partial M}{\partial x} = Q. \tag{1.3.9}$$

Equation $(1.3.9)_2$ is the so-called shear force-bending moment relation. From Eqs. $(1.3.9)_2$ and $(1.3.7)$,

$$Q = M' = -(EIv'')'. \tag{1.3.10}$$

The substitution of Eq. (1.3.10) into Eq. $(1.3.9)_1$ yields a single equation for the deflection, as follows:

$$-(EIv'')'' + F_y = \rho A \ddot{v}. \tag{1.3.11}$$

As an example, consider waves propagating in an unbounded, homogeneous and free beam $(F_y = 0)$ described by

$$v = V \exp[i(kx - \omega t)]. \tag{1.3.12}$$

The substitution of Eq. (1.3.12) into Eq. (1.3.11) yields

$$EIVk^4 = \rho A V \omega^2, \tag{1.3.13}$$

which determines the dispersion relation of the wave as

$$\omega = \sqrt{\frac{EI}{\rho A}} k^2, \tag{1.3.14}$$

or

$$\frac{\omega}{k} = \sqrt{\frac{EI}{\rho A}} k. \tag{1.3.15}$$

Equation (1.3.15) shows that the wave speed depends on the wave number k. Hence, the wave is said to be dispersive.

1.4   Plane Stress Theory of Elasticity

In conventional mechanics of materials [1], the state of plane stress for a thin film or plate is introduced for stress analysis to obtain maximal normal and shear stresses. A complete presentation on the derivation of the basic equations of plane-stress elasticity is given in this section. The plate is under in-plane loads only and undergoes in-plane extension and shear (see Fig. 1.16).

The motion and deformation of the plate is described by the following two displacement functions [4]:

$$u = u(x, y, t), \quad v = v(x, y, t). \tag{1.4.1}$$

Consider two infinitesimal and orthogonal material fibers PM and PN before deformation as shown in Fig. 1.17. After deformation, PM becomes $P'M'$. We have

$$(P'M')^2 = \left(\Delta x + \frac{\partial u}{\partial x}\Delta x\right)^2 + \left(\frac{\partial v}{\partial x}\Delta x\right)^2 \cong (\Delta x)^2 + 2(\Delta x)^2\frac{\partial u}{\partial x}, \tag{1.4.2}$$

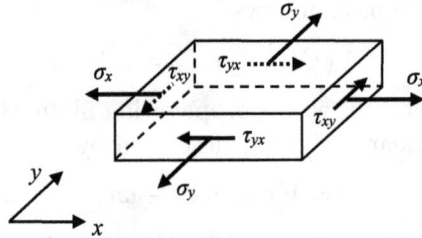

Fig. 1.16.   State of plane stress for a thin plate.

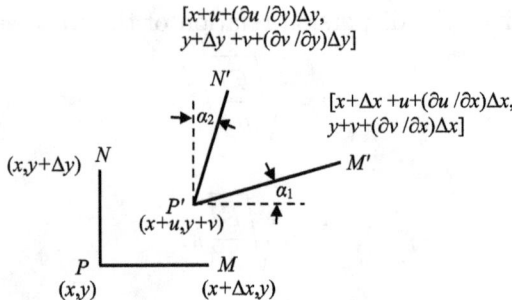

Fig. 1.17.   Displacements of two infinitesimal and orthogonal fibers.

where the approximation is based on the small displacement gradient assumption, e.g., $|\partial u/\partial x| \ll 1$. Then

$$P'M' \cong \Delta x \left(1 + 2\frac{\partial u}{\partial x}\right)^{1/2} \cong \Delta x \left(1 + \frac{\partial u}{\partial x}\right). \qquad (1.4.3)$$

Hence,

$$\varepsilon_x = \frac{P'M' - PM}{PM} = \frac{\Delta x(1 + \partial u/\partial x) - \Delta x}{\Delta x} = \frac{\partial u}{\partial x}. \qquad (1.4.4)$$

Similarly,

$$\varepsilon_y = \frac{\partial v}{\partial y}. \qquad (1.4.5)$$

From Fig. 1.17, we also have

$$\alpha_1 \cong \tan\alpha_1 = \frac{\Delta x \partial v/\partial x}{\Delta x(1 + \partial u/\partial x)} \cong \frac{\partial v}{\partial x}, \qquad (1.4.6)$$

$$\alpha_2 \cong \tan\alpha_2 = \frac{\Delta y \partial u/\partial y}{\Delta y(1 + \partial v/\partial y)} \cong \frac{\partial u}{\partial y}. \qquad (1.4.7)$$

Then the shear strain is given by

$$\gamma_{xy} = \alpha_1 + \alpha_2 \cong \frac{\partial v}{\partial x} + \frac{\partial u}{\partial y}. \qquad (1.4.8)$$

The average counterclockwise rotation of the two fibers is given by

$$\omega_z = \frac{\alpha_1 - \alpha_2}{2} \cong \frac{1}{2}\left(\frac{\partial v}{\partial x} - \frac{\partial u}{\partial y}\right). \qquad (1.4.9)$$

In plane stress, the extensional strains can be calculated from the uniaxial state of stress by superposition as

$$\varepsilon_x = \frac{1}{E}(\sigma_x - \nu\sigma_y),$$
$$\varepsilon_y = \frac{1}{E}(\sigma_y - \nu\sigma_x), \qquad (1.4.10)$$

whose inversion is given by

$$\sigma_x = \frac{E}{1 - \nu^2}(\varepsilon_x + \nu\varepsilon_y),$$

$$\sigma_y = \frac{E}{1 - \nu^2}(\varepsilon_y + \nu\varepsilon_x). \qquad (1.4.11)$$

The shear strain–stress relation is simply

$$\gamma_{xy} = \frac{\tau_{xy}}{G}, \qquad (1.4.12)$$

where G is the shear modulus and

$$G = \frac{E}{2(1 + \nu)}. \qquad (1.4.13)$$

A strain energy (internal energy) density per unit volume can be introduced as

$$U(\varepsilon_x, \varepsilon_y, \gamma_{xy}) = \frac{1}{2}(\sigma_x\varepsilon_x + \sigma_y\varepsilon_y + \tau_{xy}\gamma_{xy})$$

$$= \frac{E}{2(1 - \nu^2)}\left(\varepsilon_x^2 + \varepsilon_y^2 + 2\nu\varepsilon_x\varepsilon_y\right) + \frac{G}{2}\gamma_{xy}^2. \qquad (1.4.14)$$

Then the stress–strain relation (constitutive relation) in Eqs. (1.4.11) and (1.4.12) can be obtained from U through

$$\sigma_x = \frac{\partial U}{\partial \varepsilon_x}, \quad \sigma_y = \frac{\partial U}{\partial \varepsilon_y}, \quad \tau_{xy} = \frac{\partial U}{\partial \gamma_{xy}}. \qquad (1.4.15)$$

Based on the free-body diagram of the differential element in Fig. 1.18 in which all of the stress and body force components in the x direction are shown, we can write the equation of motion

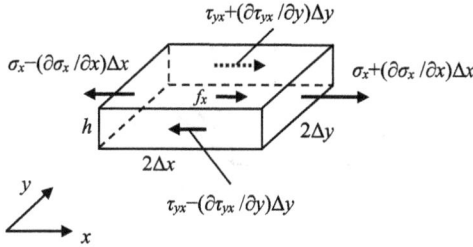

Fig. 1.18. A differential element with loads in the x direction.

(Newton's law) in the x direction as

$$\sum F_x = \left(\sigma_x + \frac{\partial \sigma_x}{\partial x}\Delta x\right)2(\Delta y)h - \left(\sigma_x - \frac{\partial \sigma_x}{\partial x}\Delta x\right)2(\Delta y)h$$

$$+ \left(\tau_{yx} + \frac{\partial \tau_{yx}}{\partial y}\Delta y\right)2(\Delta x)h - \left(\tau_{yx} - \frac{\partial \tau_{yx}}{\partial y}\Delta y\right)2(\Delta x)h$$

$$+ f_x 2(\Delta x)2(\Delta y)h = \rho 2(\Delta x)2(\Delta y)h\frac{\partial^2 u}{\partial t^2}, \qquad (1.4.16)$$

or

$$\frac{\partial \sigma_x}{\partial x} + \frac{\partial \tau_{yx}}{\partial y} + f_x = \rho\frac{\partial^2 u}{\partial t^2}. \qquad (1.4.17)$$

Similarly,

$$\frac{\partial \tau_{xy}}{\partial x} + \frac{\partial \sigma_y}{\partial y} + f_y = \rho\frac{\partial^2 v}{\partial t^2}. \qquad (1.4.18)$$

Figure 1.19 is the free-body diagram of a differential element showing all stress components with contributions to the moment equation about the z-axis which goes through the center of the element. From the moment equation

$$\sum M_z = I_z\dot{\omega}_z, \qquad (1.4.19)$$

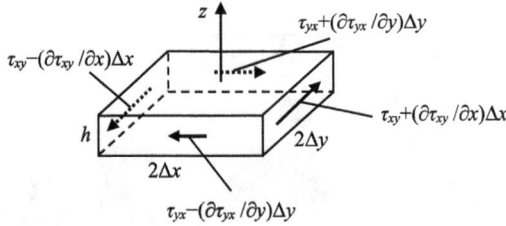

Fig. 1.19.   A differential element with loads contributing to the moment about the z-axis.

we have

$$\left(\tau_{xy} + \frac{\partial \tau_{xy}}{\partial x}\Delta x\right) 2(\Delta y)h(\Delta x) + \left(\tau_{xy} - \frac{\partial \tau_{xy}}{\partial x}\Delta x\right) 2(\Delta y)h(\Delta x)$$

$$- \left(\tau_{yx} + \frac{\partial \tau_{yx}}{\partial y}\Delta y\right) 2(\Delta x)h(\Delta y) - \left(\tau_{yx} - \frac{\partial \tau_{yx}}{\partial y}\Delta y\right) 2(\Delta x)h(\Delta y)$$

$$= I_z \dot{\omega}_z \cong 0, \qquad\qquad\qquad (1.4.20)$$

because

$$I_z = \frac{1}{12}\rho(2\Delta x)(2\Delta y)h[(2\Delta x)^2 + (2\Delta y)^2], \qquad\qquad (1.4.21)$$

is a higher-order infinitesimal. Equation (1.4.20) leads to

$$\tau_{xy} = \tau_{yx}. \qquad\qquad\qquad (1.4.22)$$

With successive substitutions from Eqs. (1.4.4), (1.4.5), (1.4.8), (1.4.11) and (1.4.12), we can write Eqs. (1.4.17) and (1.4.18) as two equations for u and v.

1.5   Three-Dimensional Theory of Elasticity

In this section, we gather the equations of the three-dimensional theory of elasticity [4] without any derivation. The physical interpretations of these equations are evident from Sec. 1.4. The notation for the positive stress components are shown in Fig. 1.20.

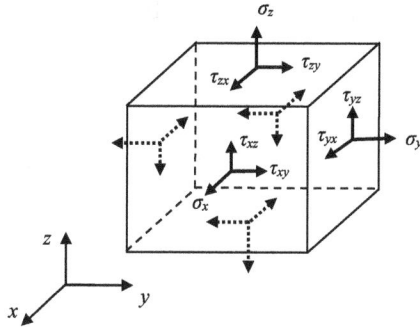

Fig. 1.20. Three-dimensional stress components.

The motion and deformation are described by the following three displacement components:

$$u = u(x, y, z, t), \quad v = v(x, y, z, t), \quad w = w(x, y, z, t). \quad (1.5.1)$$

The strains and rotations are related to the displacements through

$$\varepsilon_x = \frac{\partial u}{\partial x}, \quad \varepsilon_y = \frac{\partial v}{\partial y}, \quad \varepsilon_z = \frac{\partial w}{\partial z},$$

$$\gamma_{yz} = \gamma_{zy} = \frac{\partial w}{\partial y} + \frac{\partial v}{\partial z}, \quad \gamma_{zx} = \gamma_{xz} = \frac{\partial u}{\partial z} + \frac{\partial w}{\partial x},$$

$$\gamma_{xy} = \gamma_{yx} = \frac{\partial v}{\partial x} + \frac{\partial u}{\partial y}, \quad (1.5.2)$$

$$\omega_x = \frac{1}{2}\left(\frac{\partial w}{\partial y} - \frac{\partial v}{\partial z}\right), \quad \omega_y = \frac{1}{2}\left(\frac{\partial u}{\partial z} - \frac{\partial w}{\partial x}\right),$$

$$\omega_z = \frac{1}{2}\left(\frac{\partial v}{\partial x} - \frac{\partial u}{\partial y}\right). \quad (1.5.3)$$

The constitutive relations (generalized Hook's law) are

$$\varepsilon_x = \frac{1}{E}[\sigma_x - \nu(\sigma_y + \sigma_z)],$$

$$\varepsilon_y = \frac{1}{E}[\sigma_y - \nu(\sigma_z + \sigma_x)], \quad (1.5.4)$$

$$\varepsilon_z = \frac{1}{E}[\sigma_z - \nu(\sigma_x + \sigma_y)],$$

$$\gamma_{yz} = \frac{\tau_{yz}}{G}, \quad \gamma_{zx} = \frac{\tau_{zx}}{G}, \quad \gamma_{xy} = \frac{\tau_{xy}}{G}. \quad (1.5.5)$$

Equations (1.5.4) and (1.5.5) can also be written in terms of Lamé's elastic constants λ and μ as

$$\sigma_x = \lambda(\varepsilon_x + \varepsilon_y + \varepsilon_z) + 2\mu\varepsilon_x,$$
$$\sigma_y = \lambda(\varepsilon_x + \varepsilon_y + \varepsilon_z) + 2\mu\varepsilon_y, \qquad (1.5.6)$$
$$\sigma_z = \lambda(\varepsilon_x + \varepsilon_y + \varepsilon_z) + 2\mu\varepsilon_z,$$

$$\tau_{yz} = \mu\gamma_{yz}, \quad \tau_{zx} = \mu\gamma_{zx}, \quad \tau_{xy} = \mu\gamma_{xy}, \qquad (1.5.7)$$

where

$$\lambda = \frac{\nu E}{(1+\nu)(1-2\nu)}, \quad \mu = G. \qquad (1.5.8)$$

With the following strain energy density:

$$U = \frac{1}{2}(\sigma_x\varepsilon_x + \sigma_y\varepsilon_y + \sigma_z\varepsilon_z + \tau_{yz}\gamma_{yz} + \tau_{zx}\gamma_{zx} + \tau_{xy}\gamma_{xy})$$
$$= \frac{\lambda}{2}(\varepsilon_x + \varepsilon_y + \varepsilon_z)^2 + \mu(\varepsilon_x^2 + \varepsilon_y^2 + \varepsilon_z^2) + \frac{\mu}{2}(\gamma_{yz}^2 + \gamma_{zx}^2 + \gamma_{xy}^2), \qquad (1.5.9)$$

we have

$$\sigma_x = \frac{\partial U}{\partial \varepsilon_x}, \quad \sigma_y = \frac{\partial U}{\partial \varepsilon_y}, \quad \sigma_z = \frac{\partial U}{\partial \varepsilon_z},$$
$$\tau_{yz} = \frac{\partial U}{\partial \gamma_{yz}}, \quad \tau_{zx} = \frac{\partial U}{\partial \gamma_{zx}}, \quad \tau_{xy} = \frac{\partial U}{\partial \gamma_{xy}}. \qquad (1.5.10)$$

Newton's law and the moment equations require that

$$\frac{\partial \sigma_x}{\partial x} + \frac{\partial \tau_{yx}}{\partial y} + \frac{\partial \tau_{zx}}{\partial z} + f_x = \rho\frac{\partial^2 u}{\partial t^2},$$
$$\frac{\partial \tau_{xy}}{\partial x} + \frac{\partial \sigma_y}{\partial y} + \frac{\partial \tau_{zy}}{\partial z} + f_y = \rho\frac{\partial^2 v}{\partial t^2}, \qquad (1.5.11)$$
$$\frac{\partial \tau_{xz}}{\partial x} + \frac{\partial \tau_{yz}}{\partial y} + \frac{\partial \sigma_z}{\partial z} + f_z = \rho\frac{\partial^2 w}{\partial t^2},$$

$$\tau_{yz} = \tau_{zy}, \quad \tau_{zx} = \tau_{xz}, \quad \tau_{xy} = \tau_{yx}. \qquad (1.5.12)$$

The substitution of Eqs. (1.5.2), (1.5.6) and (1.5.7) into Eq. (1.5.11) yields three equations, for u, v and w.

As an example, consider the following one-dimensional dynamic problem described by

$$u = u(x,t), \quad v = 0, \quad w = 0. \tag{1.5.13}$$

The only strain component is

$$\varepsilon_x = \frac{\partial u}{\partial x}. \tag{1.5.14}$$

The stress components are

$$\sigma_x = (\lambda + 2\mu)\varepsilon_x = (\lambda + 2\mu)\frac{\partial u}{\partial x},$$
$$\sigma_y = \sigma_z = \lambda\varepsilon_x = \lambda\frac{\partial u}{\partial x}. \tag{1.5.15}$$

The nontrivial equation of motion is

$$\frac{\partial \sigma_x}{\partial x} = \rho\frac{\partial^2 u}{\partial t^2}. \tag{1.5.16}$$

Substituting Eq. $(1.5.15)_1$ into Eq. $(1.5.16)$, we obtain an equation for u as follows:

$$(\lambda + 2\mu)\frac{\partial^2 u}{\partial x^2} = \rho\frac{\partial^2 u}{\partial t^2}. \tag{1.5.17}$$

Consider the propagation of the following longitudinal wave:

$$u = U \exp[i(kx - \omega t)]. \tag{1.5.18}$$

The substitution of Eq. $(1.5.18)$ into Eq. $(1.5.17)$ leads to the following longitudinal wave speed:

$$\frac{\omega}{k} = \sqrt{\frac{\lambda + 2\mu}{\rho}} = c_1, \tag{1.5.19}$$

where

$$\lambda + 2\mu = \frac{E(1-v)}{(1+v)(1-2v)}. \tag{1.5.20}$$

c_1 is different from the longitudinal wave speed in extensional motions of a thin rod in Eq. (1.1.10).

As another example, consider motions described by

$$u = 0, \quad v = v(x,t), \quad w = 0. \tag{1.5.21}$$

The only strain component is

$$\gamma_{xy} = \frac{\partial v}{\partial x}. \tag{1.5.22}$$

The corresponding stress component is

$$\tau_{xy} = \mu \gamma_{xy} = \mu \frac{\partial v}{\partial x}. \tag{1.5.23}$$

The only nontrivial equation of motion takes the following form:

$$\mu \frac{\partial^2 v}{\partial x^2} = \rho \frac{\partial^2 v}{\partial t^2}. \tag{1.5.24}$$

Consider the propagation of the following transverse or shear wave:

$$v = V \exp[i(kx - \omega t)]. \tag{1.5.25}$$

Substituting Eq. (1.5.25) into Eq. (1.5.24), we obtain the transverse wave speed as

$$\frac{\omega}{k} = \sqrt{\frac{\mu}{\rho}} = c_2. \tag{1.5.26}$$

1.6 Indicial Notation and Cartesian Tensors

To write the three-dimensional equations of elasticity more efficiently, we introduced the so-called indicial notation [5]. The components of the position vector $\mathbf{x} = x\mathbf{i} + y\mathbf{j} + z\mathbf{k}$ of a point is written as x_i with i assuming 1, 2 or 3, i.e.,

$$(x, y, z) = (x_1, x_2, x_3). \tag{1.6.1}$$

The index i is called a free index. x_k represents the same position vector as x_i. Similarly, the components of the displacement vector $\mathbf{u}$

are written as u_k according to

$$(u, v, w) = (u_x, u_y, u_z) = (u_1, u_2, u_3). \tag{1.6.2}$$

The strain components are represented by a two-dimensional array as follows:

$$\begin{bmatrix} \varepsilon_x & \gamma_{xy}/2 & \gamma_{xz}/2 \\ \gamma_{yx}/2 & \varepsilon_y & \gamma_{yz}/2 \\ \gamma_{zx}/2 & \gamma_{zy}/2 & \varepsilon_z \end{bmatrix} = \begin{bmatrix} S_{11} & S_{12} & S_{13} \\ S_{21} & S_{22} & S_{23} \\ S_{31} & S_{32} & S_{33} \end{bmatrix}, \tag{1.6.3}$$

where S_{kl} with k and l assuming 1, 2 or 3 is symmetric, i.e., $S_{kl} = S_{lk}$. S_{kl} is called the strain tensor (a second-order tensor) because it transforms in a specific manner under a rotation of the coordinate system (x_1, x_2, x_3). The three rotations are written as

$$\omega_1 = \frac{1}{2}\left(\frac{\partial u_3}{\partial x_2} - \frac{\partial u_2}{\partial x_3}\right), \quad \omega_2 = \frac{1}{2}\left(\frac{\partial u_1}{\partial x_3} - \frac{\partial u_3}{\partial x_1}\right),$$

$$\omega_3 = \frac{1}{2}\left(\frac{\partial u_2}{\partial x_1} - \frac{\partial u_1}{\partial x_2}\right). \tag{1.6.4}$$

The stress components are represented by the stress tensor T_{kl} according to

$$\begin{bmatrix} \sigma_x & \tau_{xy} & \tau_{xz} \\ \tau_{yx} & \sigma_y & \tau_{yz} \\ \tau_{zx} & \tau_{zy} & \sigma_z \end{bmatrix} = \begin{bmatrix} T_{11} & T_{12} & T_{13} \\ T_{21} & T_{22} & T_{23} \\ T_{31} & T_{32} & T_{33} \end{bmatrix}, \tag{1.6.5}$$

which is also symmetric.

To simplify the notation further, we introduce a summation convention that repeated indices (called dummy indices) are summed from 1 to 3. For example,

$$S_{kk} = \sum_{k=1}^{3} S_{kk} = S_{11} + S_{22} + S_{33} = \sum_{l=1}^{3} S_{ll} = S_{ll}. \tag{1.6.6}$$

With the summation convention, the position vector of a point can be written as $\mathbf{x} = x_i \mathbf{e}_i$ where $\mathbf{e}_1 = \mathbf{i}$, $\mathbf{e}_2 = \mathbf{j}$ and $\mathbf{e}_3 = \mathbf{k}$. Similarly,

$\mathbf{u} = u_j \mathbf{e}_j$. The definition of the multiplication of two 3×3 matrices $[A][B] = [C]$ can be written as

$$C_{ij} = \sum_{k=1}^{3} A_{ik} B_{kj} = A_{ik} B_{kj} = A_{il} B_{lj}. \qquad (1.6.7)$$

We also introduce the so-called comma convention for partial differentiation with respect to a coordinate

$$\frac{\partial(\ )}{\partial x_k} = (\ )_{,k}. \qquad (1.6.8)$$

Then the divergence of a vector field $\mathbf{u}$ can be written as

$$\nabla \cdot \mathbf{u} = \frac{\partial u_1}{\partial x_1} + \frac{\partial u_2}{\partial x_2} + \frac{\partial u_3}{\partial x_3} = u_{k,k} = u_{l,l}. \qquad (1.6.9)$$

The strain-displacement relation can be written as

$$S_{ij} = (u_{j,i} + u_{i,j})/2 = S_{ji}. \qquad (1.6.10)$$

The second-order unit or identity tensor is called the Kronecker delta and is defined by

$$[\delta_{kl}] = \begin{bmatrix} 1 & 0 & 0 \\ 0 & 1 & 0 \\ 0 & 0 & 1 \end{bmatrix}, \qquad (1.6.11)$$

which, when multiplied to a vector or another second-order tensor, produces the same vector or tensor. The stress–strain relation can be written as

$$T_{ij} = \lambda S_{kk} \delta_{ij} + 2\mu S_{ij}, \qquad (1.6.12)$$

or

$$S_{ij} = \frac{1}{E}[(1 + \nu)T_{ij} - \nu T_{kk} \delta_{ij}]. \qquad (1.6.13)$$

The equation of motion and the moment equations take the following form:

$$T_{ji,j} + f_i = \rho \ddot{u}_i, \qquad (1.6.14)$$

$$T_{ij} = T_{ji}. \qquad (1.6.15)$$

We note that Eq. (1.6.15) is in fact implied by Eq. (1.6.12).

The strain energy density can be written as

$$U = \frac{1}{2}T_{ij}S_{ij} = \frac{1}{2}c_{ijkl}S_{ij}S_{kl}, \tag{1.6.16}$$

where, for isotropic materials,

$$c_{ijkl} = \lambda\delta_{ij}\delta_{kl} + \mu(\delta_{ik}\delta_{jl} + \delta_{il}\delta_{jk}). \tag{1.6.17}$$

c_{ijkl} is a fourth-order tensor called the elastic stiffness. The stress–strain relation can be obtained from U through

$$T_{kl} = \frac{1}{2}\left(\frac{\partial U}{\partial S_{kl}} + \frac{\partial U}{\partial S_{lk}}\right) = \frac{\partial U}{\partial S_{kl}}, \tag{1.6.18}$$

or

$$T_{ij} = c_{ijkl}S_{kl} = \sum_{k=1}^{3}\sum_{l=1}^{3}c_{ijkl}S_{kl}. \tag{1.6.19}$$

In general, the stiffness tensor has the following symmetries:

$$c_{ijkl} = c_{jikl} = c_{klij}. \tag{1.6.20}$$

The inverse of Eq. (1.6.19) is

$$S_{ij} = s_{ijkl}T_{kl}, \tag{1.6.21}$$

where s_{ijkl} is the elastic compliance tensor.

1.7 Matrix Notation for Anisotropic Elasticity

We now introduce a compact matrix notation [5]. It will allow us to use matrices to represent the stress, strain, elastic stiffness and compliance tensors and will be convenient for constitutive relations of anisotropic materials. This notation consists of replacing pairs of tensor indices ij or kl by single matrix indices p or q, where i, j, k

and l take the values of 1, 2 and 3, and p and q take the values of 1, 2, 3, 4, 5 and 6 according to

$$ij \text{ or } kl: \quad 11 \quad 22 \quad 33 \quad 23 \text{ or } 32 \quad 31 \text{ or } 13 \quad 12 \text{ or } 21$$

$$p \text{ or } q: \quad 1 \quad 2 \quad 3 \quad 4 \quad 5 \quad 6 \tag{1.7.1}$$

Thus, for the stress tensor, we have

$$T_1 = T_{11}, \quad T_2 = T_{22}, \quad T_3 = T_{33},$$
$$T_4 = T_{23}, \quad T_5 = T_{31}, \quad T_6 = T_{12}. \tag{1.7.2}$$

For the strain tensor, we introduce S_p such that

$$S_1 = S_{11}, \quad S_2 = S_{22}, \quad S_3 = S_{33},$$
$$S_4 = 2S_{23}, \quad S_5 = 2S_{31}, \quad S_6 = 2S_{12}. \tag{1.7.3}$$

Accordingly, the strain energy density can be written as

$$U = \frac{1}{2} T_p S_p = \frac{1}{2} c_{pq} S_p S_q. \tag{1.7.4}$$

The constitutive relations take the following form:

$$T_p = \frac{\partial U}{\partial S_p} = c_{pq} S_q, \quad c_{pq} = c_{qp},$$
$$S_p = s_{pq} T_q, \quad s_{pq} = s_{qp}. \tag{1.7.5}$$

The elastic properties of a fully anisotropic material such as a triclinic crystal are represented by the following full array of $[c_{pq}]$ with 21 independent material constants:

$$
\begin{bmatrix} T_1 \\ T_2 \\ T_3 \\ T_4 \\ T_5 \\ T_6 \end{bmatrix}
=
\begin{bmatrix}
c_{11} & c_{12} & c_{13} & c_{14} & c_{15} & c_{16} \\
c_{21} & c_{22} & c_{23} & c_{24} & c_{25} & c_{26} \\
c_{31} & c_{32} & c_{33} & c_{34} & c_{35} & c_{36} \\
c_{41} & c_{42} & c_{43} & c_{44} & c_{45} & c_{46} \\
c_{51} & c_{52} & c_{53} & c_{54} & c_{55} & c_{56} \\
c_{61} & c_{62} & c_{63} & c_{64} & c_{65} & c_{66}
\end{bmatrix}
\begin{bmatrix} S_1 \\ S_2 \\ S_3 \\ S_4 \\ S_5 \\ S_6 \end{bmatrix}.
\tag{1.7.6}
$$

Table 1.1.   Elastic constants of isotropic materials.

	c_{11}, c_{12}	λ, μ	E, v
c_{11}	c_{11}	$\lambda + 2\mu$	$\dfrac{E(1-v)}{(1+v)(1-2v)}$
c_{12}	c_{12}	λ	$\dfrac{Ev}{(1+v)(1-2v)}$
λ	c_{12}	λ	$\dfrac{Ev}{(1+v)(1-2v)}$
μ	$\dfrac{c_{11}-c_{12}}{2}$	μ	$\dfrac{E}{2(1+v)}$
E	$\dfrac{(c_{11}+2c_{12})(c_{11}-c_{12})}{c_{11}+c_{12}}$	$\dfrac{\mu(3\lambda+2\mu)}{\lambda+\mu}$	E
v	$\dfrac{c_{12}}{c_{11}+c_{12}}$	$\dfrac{\lambda}{2(\lambda+\mu)}$	v

In the special case of an isotropic material with two independent material constants, the stress–strain relation reduces to

$$T_1 = c_{11}S_1 + c_{12}S_2 + c_{12}S_3,$$
$$T_2 = c_{21}S_1 + c_{11}S_2 + c_{12}S_3, \qquad (1.7.7)$$
$$T_3 = c_{21}S_1 + c_{21}S_2 + c_{11}S_3,$$
$$T_4 = c_{44}S_4, \quad T_5 = c_{44}S_5, \quad T_6 = c_{44}S_6,$$

where

$$c_{44} = \frac{1}{2}(c_{11} - c_{12}). \qquad (1.7.8)$$

The relations of the constants c_{11} and c_{12} to Lamé's constants (λ, μ), to Young's modulus, E, and Poisson's ratio, v, are given in Table 1.1 [5].

1.8   Timoshenko Theory for Bending of Beams

In this section, we generalize the Euler–Bernoulli theory for bending of beams in Sec. 1.3 to the Timoshenko theory [6] for bending by including the effects of shear deformation and rotatory inertia which are significant when a beam is not very long or thin. Different

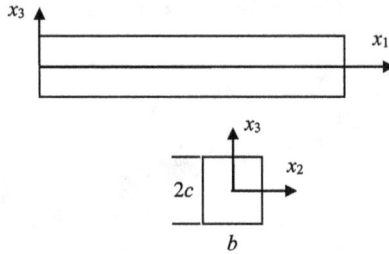

Fig. 1.21.   Side view and cross-section of a beam.

from Secs. 1.1–1.3 and [1,6] where the one-dimensional theories are constructed directly without the use of the three-dimensional theory of elasticity, in this section we derive one-dimensional equations from the three-dimensional equations through a systematic procedure developed by Mindlin [7]. The procedure does not need Fig. 1.13 and it produces expressions of the effective one-dimensional material constants in terms of the three-dimensional material constants. Consider the beam in Fig. 1.21. It is in bending with shear deformation in the (x_1, x_3) plane. For simplicity, we consider a beam with a rectangular cross-section. The procedure in [7] is valid for other cross-sections in general.

We begin with the following approximations of the relevant displacement components:

$$
\begin{aligned}
u_3(\mathbf{x}, t) &\cong w(x_1, t), \\
u_1(\mathbf{x}, t) &\cong x_3 \psi(x_1, t),
\end{aligned}
\tag{1.8.1}
$$

where $w(x_1, t)$ is the bending displacement (deflection) and $\psi(x_1, t)$, the shear displacement. The relevant strain components are calculated from the three-dimensional equations as

$$
\begin{aligned}
S_1 &= S_{11} = u_{1,1} = x_3 \psi_{,1}, \\
S_5 &= 2S_{31} = u_{3,1} + u_{1,3} = w_{,1} + \psi.
\end{aligned}
\tag{1.8.2}
$$

It can be seen that w and ψ together contribute to the shear strain S_5.

We limit ourselves to materials without couplings between extensional and shear deformations, and interactions among different shear strains. The material matrices are assumed to be

$$
\begin{bmatrix}
c_{11} & c_{12} & c_{13} & 0 & 0 & 0 \\
c_{21} & c_{22} & c_{23} & 0 & 0 & 0 \\
c_{31} & c_{32} & c_{33} & 0 & 0 & 0 \\
0 & 0 & 0 & c_{44} & 0 & 0 \\
0 & 0 & 0 & 0 & c_{55} & 0 \\
0 & 0 & 0 & 0 & 0 & c_{66}
\end{bmatrix},
\begin{bmatrix}
s_{11} & s_{12} & s_{13} & 0 & 0 & 0 \\
s_{21} & s_{22} & s_{23} & 0 & 0 & 0 \\
s_{31} & s_{32} & s_{33} & 0 & 0 & 0 \\
0 & 0 & 0 & s_{44} & 0 & 0 \\
0 & 0 & 0 & 0 & s_{55} & 0 \\
0 & 0 & 0 & 0 & 0 & s_{66}
\end{bmatrix}.
$$

$$(1.8.3)$$

For bending in the (x_1, x_3) plane, the main stress components over a cross-section are the normal stress T_1 and shear stress $T_{13} = T_5$. We introduce the following stress relaxation [7] for thin beams which allows the development of S_2 and S_3 to account for Poisson's effect:

$$T_2 = T_3 = 0. \tag{1.8.4}$$

In this case, it is convenient to use the elastic compliance. The relevant constitutive relations are

$$S_1 = s_{11}T_1 + s_{12}T_2 + s_{13}T_3 = s_{11}T_1,$$
$$S_5 = s_{55}T_5. \tag{1.8.5}$$

We invert Eq. (1.8.5) for expressions of stresses in terms of strains

$$T_1 = T_{11} = \bar{c}_{11}S_1,$$
$$T_5 = T_{13} = \bar{c}_{55}S_5, \tag{1.8.6}$$

where the effective one-dimensional material constants for thin beams are denoted by

$$\bar{c}_{11} = 1/s_{11}, \quad \bar{c}_{55} = 1/s_{55}. \tag{1.8.7}$$

For isotropic materials, it can be shown that $1/s_{11} = E$. Alternatively, if the elastic stiffness matrix is used, we have

$$T_1 = c_{11}S_1 + c_{12}S_2 + c_{13}S_3,$$
$$T_2 = c_{21}S_1 + c_{22}S_2 + c_{23}S_3 = 0,$$
$$T_3 = c_{31}S_1 + c_{32}S_2 + c_{33}S_3 = 0,$$
$$T_5 = c_{55}S_5. \tag{1.8.8}$$

We then invert Eq. $(1.8.8)_{2,3}$ for expressions of S_2 and S_3 in terms of S_1, and substitute the expressions into Eq. $(1.8.8)_1$. This yields the following effective one-dimensional material constants needed in the one-dimensional constitutive relations in Eq. $(1.8.6)$:

$$\bar{c}_{11} = c_{11} + \frac{2c_{23}c_{31}c_{12} - c_{33}c_{12}^2 - c_{22}c_{13}^2}{c_{22}c_{33} - c_{23}^2}, \tag{1.8.9}$$

$$\bar{c}_{55} = c_{55},$$

which can be shown to be equal to those in Eq. $(1.8.7)$. With substitution from Eq. $(1.8.2)$, the one-dimensional constitutive relation in Eq. $(1.8.6)$ becomes

$$T_1 = \bar{c}_{11}S_1 = \bar{c}_{11}x_3\psi_{,1},$$
$$T_5 = \bar{c}_{55}S_5 = \bar{c}_{55}(w_{,1} + \psi). \tag{1.8.10}$$

The transverse shear force Q and bending moment M over a cross-section of the beam are defined by the following integrals:

$$Q = \lambda \int_A T_{13}\, dx_2\, dx_3 = \lambda \bar{c}_{55}A(w_{,1} + \psi),$$

$$M = \int_A x_3 T_1\, dx_2\, dx_3 = \bar{c}_{11}I\psi_{,1}, \tag{1.8.11}$$

where

$$A = 2bc, \quad I = \frac{2}{3}bc^3. \tag{1.8.12}$$

λ is a correction factor introduced to improve the accuracy of the theory. Its value depends on the shape of the cross-section. This correction factor will be dropped in the rest of the book. It can be included and determined whenever needed.

From the equation of motion of the differential element of the beam in Fig. 1.22 in the x_3 direction and its moment equation, we obtain

$$Q_{,1} + F_3 = \rho A \ddot{w}, \tag{1.8.13}$$

$$M_{,1} - Q + m_2 = \rho I \ddot{w}_2 = \rho I \ddot{\psi}, \tag{1.8.14}$$

where $F_3(x_1, t)$ is the transverse force per unit length of the beam, and $m_2(x_1, t)$, the distributed moment per unit length. In obtaining

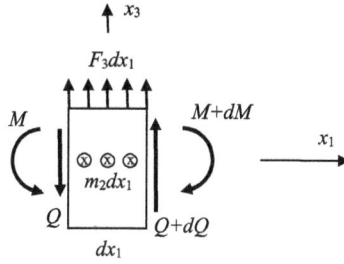

Fig. 1.22. A differential element of the beam under mechanical loads.

the moment equation in Eq. (1.8.14), the rotation about the x_2-axis has been approximated by

$$\omega_2 = \frac{1}{2}\left(\frac{\partial u_1}{\partial x_3} - \frac{\partial u_3}{\partial x_1}\right) = \frac{1}{2}(\psi - w_{,1}) \cong \frac{1}{2}(\psi + \psi) = \psi. \quad (1.8.15)$$

When calculating ω_2, we have set $S_5 = 0$ so that $w_{,1} \cong -\psi$ from Eq. (1.8.2). The substitution of Eq. (1.8.11) into Eqs. (1.8.13) and (1.8.14) gives two equations, for w and ψ, as follows:

$$[\lambda \bar{c}_{55}A(w_{,1} + \psi)]_{,1} + F_3 = \rho A \ddot{w},$$

$$(\bar{c}_{11}I\psi_{,1})_{,1} - \lambda \bar{c}_{55}A(w_{,1} + \psi) + m_2 = \rho I \ddot{\psi}. \quad (1.8.16)$$

We note that the above derivation is somewhat mixed in the sense that Eqs. (1.8.13) and (1.8.14) are derived using a one-dimensional differential element of the beam. In fact these equations can also be derived from the three-dimensional equations of motion in elasticity without the need of Fig. 1.22 [7].

The Euler–Bernoulli theory can be reduced from the above equations as a special case when the shear deformation and rotatory inertia are negligible:

$$S_5 = w_{,1} + \psi \cong 0, \quad I \cong 0. \quad (1.8.17)$$

Then, Eq. (1.8.14) reduces to the following shear force-bending moment relation which serves as the constitutive relation for the shear force:

$$Q = M_{,1} + m_2 = \bar{c}_{11}I\psi_{,11} + m_2 = -\bar{c}_{11}Iw_{,111} + m_2. \quad (1.8.18)$$

Substituting Eq. (1.8.18) into Eq. (1.8.13), we recover the equation for bending of the Euler–Bernoulli theory in Eq. (1.3.11) as follows:

$$-\bar{c}_{11}Iw_{,1111} + m_{2,1} + F_3 = \rho A \ddot{w}. \quad (1.8.19)$$

Chapter 2

Heat Conduction and Thermoelasticity

This chapter begins with one-dimensional heat conduction in a rigid rod and then extends it to three-dimensional heat conduction in rigid solids. These are followed by the theory of thermoelasticity which treats coupled heat conduction and elastic deformation. One-dimensional theories for the extension and bending of thermoelastic beams are also derived.

2.1 One-Dimensional Heat Conduction

Consider heat conduction along the axial (x) direction of the rigid rod in Fig. 2.1. Its cross-sectional area is A.

Let $T(x,t)$ be the absolute temperature along the rod. Θ_0 is a uniform reference temperature. $\theta(x,t) = T - \Theta_0$ is a small temperature deviation from Θ_0. The axial heat flux density per unit area and unit time is denoted by $h(x,t)$. The heat source per unit volume and unit time is $r(x,t)$. The specific heat at Θ_0 is C_V. Figure 2.2 shows a differential element of the rod. During an infinitesimal time interval dt, the conservation of energy (the first law of thermodynamics) of the element takes the following form:

$$hA\,dt - [hA + d(hA)]dt + rA\,dx\,dt$$
$$= C_V A\,dx[\theta(t+dt) - \theta(t)], \qquad (2.1.1)$$

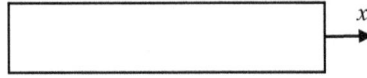

Fig. 2.1.   A rigid rod along x.

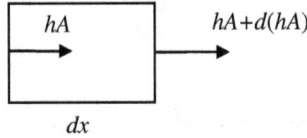

Fig. 2.2.   A differential element of the rod under thermal loads.

or

$$-\frac{\partial(hA)}{\partial x} + rA = C_V A \frac{\partial \theta}{\partial t}. \tag{2.1.2}$$

In addition, for the heat flux we have Fourier's law

$$h = -\kappa \frac{\partial \theta}{\partial x}, \tag{2.1.3}$$

where κ is an effective one-dimensional heat conduction coefficient. The substitution of Eq. (2.1.3) into Eq. (2.1.2) yields a single equation for $\theta(x,t)$, as follows:

$$A C_V \frac{\partial \theta}{\partial t} = Ar + \frac{\partial}{\partial x}\left(A\kappa \frac{\partial \theta}{\partial x}\right). \tag{2.1.4}$$

For a homogeneous rod with a constant A and a constant κ, Eq. (2.1.4) reduces to

$$C_V \frac{\partial \theta}{\partial t} = \kappa \frac{\partial^2 \theta}{\partial x^2} + r. \tag{2.1.5}$$

As an example, consider the static state of a finite and source-free rod with specified temperatures at its two ends. The boundary-value problem is

$$\kappa A \theta'' = 0, \quad 0 < x < L,$$
$$\theta(0) = a, \quad \theta(L) = b. \tag{2.1.6}$$

The solution is a linear temperature distribution given by

$$\theta = a + \frac{b-a}{L}x. \tag{2.1.7}$$

2.2 Three-Dimensional Heat Conduction

In this section, we generalize the one-dimensional equations in the previous section to the three-dimensional situation without any derivation. The physical interpretations of the equations are self-evident based on the understanding of the previous section. The conservation of energy takes the form of

$$-h_{k,k} + r = C_V \frac{\partial \theta}{\partial t}, \tag{2.2.1}$$

where h_k is the heat flux vector and r, the body heat source. The three-dimensional version of Fourier's law is given by

$$h_k = -\kappa_{kl}\theta_{,l}, \tag{2.2.2}$$

or

$$\begin{bmatrix} h_1 \\ h_2 \\ h_3 \end{bmatrix} = - \begin{bmatrix} \kappa_{11} & \kappa_{12} & \kappa_{13} \\ \kappa_{21} & \kappa_{22} & \kappa_{23} \\ \kappa_{31} & \kappa_{32} & \kappa_{33} \end{bmatrix} \begin{bmatrix} \theta_{,1} \\ \theta_{,2} \\ \theta_{,3} \end{bmatrix}, \tag{2.2.3}$$

where κ_{kl} is the heat conduction coefficient tensor. In Sec. 2.4, κ_{kl} will be shown to be symmetric and positive definite with the use of the second law of thermodynamics. The substitution of Eq. (2.2.2) into Eq. (2.2.1) yields an equation for θ as follows:

$$C_V \frac{\partial \theta}{\partial t} = (\kappa_{kl}\theta_{,l})_{,k} + r. \tag{2.2.4}$$

As an example, consider the static state of a finite and homogeneous body V under a uniform temperature θ_0 on its boundary S. The boundary-value problem is

$$\kappa_{kl}\theta_{,kl} = 0, \quad \text{in } V,$$
$$\theta = \theta_0, \quad \text{on } S. \tag{2.2.5}$$

The solution is a uniform temperature inside the body:

$$\theta \equiv \theta_0, \quad \text{in } V. \tag{2.2.6}$$

2.3 One-Dimensional Thermoelasticity

When the rod under thermal loads in Figs. 2.1 and 2.2 is not rigid, thermoelastic interactions occur. Consider the differential element of a thermoelastic rod shown in Fig. 2.3. The axial direction is x_1. In addition to the thermal loads in Fig. 2.2, the stresses and velocities of the two cross-sections on the left and right are shown in Fig. 2.3. The element is also under distributed body force f_1 per unit volume in the x_1 direction. For simplicity, we consider a homogeneous rod with a constant cross-section and uniform material properties.

Newton's second law takes the following form:

$$(T_{11} + dT_{11})A - T_{11}A + f_1 A\, dx_1 = \rho(A\, dx_1)\dot{v}_1, \tag{2.3.1}$$

or

$$\frac{\partial T_{11}}{\partial x_1} + f_1 = \rho \ddot{u}_1. \tag{2.3.2}$$

The conservation of energy in rate (power) form can be stated as

$$\frac{\partial}{\partial t}\left(UA\, dx_1 + \frac{1}{2}\rho v_1^2 A\, dx_1 \right)$$
$$= (T_{11} + dT_{11})A(v_1 + dv_1) - T_{11}Av_1 + f_1 v_1 A\, dx_1$$
$$+ rA\, dx_1 + h_1 A - [h_1 A + d(h_1 A)], \tag{2.3.3}$$

$$\frac{\partial U}{\partial t}A\, dx_1 + \rho v_1 \dot{v}_1 A\, dx_1$$
$$= T_{11}A\frac{\partial v_1}{\partial x_1}dx_1 + Av_1\frac{\partial T_{11}}{\partial x_1}dx_1 + f_1 v_1 A\, dx_1$$
$$+ rA\, dx_1 - \frac{\partial h_1}{\partial x_1}A\, dx_1, \tag{2.3.4}$$

$$\frac{\partial U}{\partial t} + v_1\left(\rho\dot{v}_1 - \frac{\partial T_{11}}{\partial x_1} - f_1 \right) = T_{11}\frac{\partial v_1}{\partial x_1} + r - \frac{\partial h_1}{\partial x_1}. \tag{2.3.5}$$

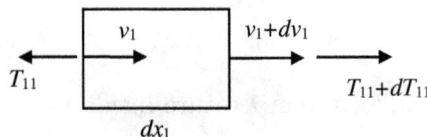

Fig. 2.3. A differential element of a thermoelastic rod under mechanical loads.

With the use of Eq. (2.3.2), the energy equation in Eq. (2.3.5) becomes

$$\frac{\partial U}{\partial t} = T_{11}\frac{\partial v_1}{\partial x_1} + r - \frac{\partial h_1}{\partial x_1}. \tag{2.3.6}$$

Let the entropy density be $\eta(x,t)$. The second law of thermodynamics for the differential element in Fig. 2.3 (or Fig. 2.2) can be written as

$$\frac{\partial}{\partial t}(\eta A\, dx_1) \geq \frac{r}{T}A\, dx_1 + \frac{h_1 A}{T} - \left[\frac{h_1 A}{T} + d\left(\frac{h_1 A}{T}\right)\right], \tag{2.3.7}$$

or

$$\frac{\partial \eta}{\partial t} \geq \frac{r}{T} - \frac{\partial}{\partial x_1}\left(\frac{h_1}{T}\right). \tag{2.3.8}$$

Eliminating r from Eqs. (2.3.6) and (2.3.8), we obtain the following Clausius–Duhem inequality:

$$T\frac{\partial \eta}{\partial t} - \frac{\partial U}{\partial t} + T_{11}\frac{\partial v_1}{\partial x_1} - \frac{h_1}{T}\frac{\partial T}{\partial x_1} \geq 0. \tag{2.3.9}$$

The Helmholtz free energy density F can be introduced through the following Legendre transform from the internal energy U:

$$F(S_{11}, T) = U(S_{11}, \eta) - T\eta. \tag{2.3.10}$$

Then

$$U = F + T\eta, \tag{2.3.11}$$

$$\dot{U} = \dot{F} + \dot{T}\eta + T\dot{\eta}, \tag{2.3.12}$$

$$T\dot{\eta} - \dot{U} = -\dot{F} - \dot{T}\eta. \tag{2.3.13}$$

The substitution of Eq. (2.3.13) into the energy equation in Eq. (2.3.6) and the Clausius–Duhem inequality in Eq. (2.3.9) leads to

$$\dot{F} + \dot{T}\eta + T\dot{\eta} = T_{11}\frac{\partial v_1}{\partial x_1} + r - \frac{\partial h_1}{\partial x_1}, \qquad (2.3.14)$$

$$-(\dot{F} + \dot{T}\eta) + T_{11}\frac{\partial v_1}{\partial x_1} - \frac{h_1}{T}\frac{\partial T}{\partial x_1} \geq 0. \qquad (2.3.15)$$

With

$$F = F(S_{11}, T), \qquad (2.3.16)$$

we have

$$\dot{F} = \frac{\partial F}{\partial S_{11}}\dot{S}_{11} + \frac{\partial F}{\partial T}\dot{T}. \qquad (2.3.17)$$

Substituting Eq. (2.3.17) into Eqs. (2.3.14) and (2.3.15), we have

$$\frac{\partial F}{\partial S_{11}}\dot{S}_{11} + \frac{\partial F}{\partial T}\dot{T} + \dot{T}\eta + T\dot{\eta} = T_{11}\frac{\partial v_1}{\partial x_1} + r - \frac{\partial h_1}{\partial x_1}, \qquad (2.3.18)$$

$$-\frac{\partial F}{\partial S_{11}}\dot{S}_{11} - \frac{\partial F}{\partial T}\dot{T} - \dot{T}\eta + T_{11}\frac{\partial v_1}{\partial x_1} - \frac{h_1}{T}\frac{\partial T}{\partial x_1} \geq 0. \qquad (2.3.19)$$

Equation (2.3.19) can be written as

$$-\left(\frac{\partial F}{\partial T} + \eta\right)\dot{T} + \left(T_{11} - \frac{\partial F}{\partial S_{11}}\right)\dot{S}_{11} - \frac{h_1}{T}\frac{\partial T}{\partial x_1} \geq 0. \qquad (2.3.20)$$

For Eq. (2.3.20) to hold for any $\dot{T}$ and any $\dot{S}_{11}$,

$$\eta = -\frac{\partial F}{\partial T}, \quad T_{11} = \frac{\partial F}{\partial S_{11}}. \qquad (2.3.21)$$

Then Eq. (2.3.20) reduces to

$$-\frac{h_1}{T}\frac{\partial T}{\partial x_1} \geq 0. \qquad (2.3.22)$$

With Eq. (2.3.21), the energy equation in Eq. (2.3.18) becomes

$$T\dot{\eta} = r - \frac{\partial h_1}{\partial x_1}. \qquad (2.3.23)$$

We expand F at Θ_0 as

$$F = F_0 - \eta_0\theta + T_{11}^0 S_{11} + \frac{1}{2}cS_{11}^2 - \lambda\theta S_{11} - \frac{1}{2}\frac{C_V}{\Theta_0}\theta^2. \qquad (2.3.24)$$

From Eqs. (2.3.24) and (2.3.21), we obtain the thermoelastic constitutive relations as

$$\eta = -\frac{\partial F}{\partial \theta} = \eta_0 + \frac{C_V}{\Theta_0}\theta + \lambda S_{11},$$
$$T_{11} = \frac{\partial F}{\partial S_{11}} = T_{11}^0 + cS_{11} - \lambda\theta. \qquad (2.3.25)$$

η^0 is immaterial and will be dropped. T_{11}^0 represents initial stress which will not be considered and is set to zero. λ is an effective one-dimensional thermoelastic constant. We also have Fourier's law

$$h_1 = -\kappa\frac{\partial\theta}{\partial x_1}, \qquad (2.3.26)$$

which, when substituted into Eq. (2.3.22), requires that

$$-\frac{h_1}{T}\frac{\partial T}{\partial x_1} = \frac{\kappa}{T}\frac{\partial\theta}{\partial x_1}\frac{\partial\theta}{\partial x_1} \geq 0. \qquad (2.3.27)$$

Hence, $\kappa > 0$.

In summary, for a linear theory, the one-dimensional equations for a thermoelastic rod along x_1 are

$$T_{11,1} + f_1 = \rho\ddot{u}_1, \qquad (2.3.28)$$
$$\Theta_0\dot{\eta} = r - \frac{\partial h_1}{\partial x_1},$$

$$T_{11} = cS_{11} - \lambda\theta,$$
$$\eta = \frac{C_V}{\Theta_0}\theta + \lambda S_{11}, \qquad (2.3.29)$$

$$h_1 = -\kappa\frac{\partial\theta}{\partial x_1}, \qquad (2.3.30)$$

$$S_{11} = u_{1,1}. \qquad (2.3.31)$$

With successive substitutions from Eqs. (2.3.29)–(2.3.31), we can write Eq. (2.3.28) as two equations for u_1 and θ, as follows:

$$cu_{1,11} - \lambda\theta_{,1} + f_1 = \rho\ddot{u}_1,$$

$$C_V\dot{\theta} + \Theta_0\lambda\dot{u}_{1,1} = \frac{\partial}{\partial x_1}\left(\kappa\frac{\partial\theta}{\partial x_1}\right) + r. \qquad (2.3.32)$$

In the special case when there is no thermoelastic coupling, Eq. (2.3.32) reduces to two uncoupled problems of one-dimensional extension of an elastic rod and one-dimensional heat conduction:

$$cu_{1,11} + f_1 = \rho\ddot{u}_1,$$

$$C_V\dot{\theta} = \frac{\partial}{\partial x_1}\left(\kappa\frac{\partial\theta}{\partial x_1}\right) + r. \qquad (2.3.33)$$

As an example, consider the static extensional deformation of a mechanically free rod within $0 < x < L$ under the same temperature change θ_0 at its two ends. The boundary-value problem is

$$cu_{1,11} = 0, \quad 0 = \frac{\partial}{\partial x_1}\left(\kappa\frac{\partial\theta}{\partial x_1}\right), \quad 0 < x < L,$$

$$T_1(0) = T_1(L) = 0, \quad \theta(0) = \theta(L) = \theta_0. \qquad (2.3.34)$$

In this case, the temperature field is uncoupled to the mechanical fields and it is found to be uniform along the rod with $\theta = \theta_0$. The extensional strain is determined from

$$T_{11} = cS_{11} - \lambda\theta = 0 \qquad (2.3.35)$$

as

$$S_{11} = \frac{\lambda}{c}\theta_0 = \alpha\theta_0, \qquad (2.3.36)$$

where α is the effective one-dimensional coefficient of thermal expansion.

2.4   Three-Dimensional Thermoelasticity

This section summarizes the three-dimensional equations of thermoelasticity [8–10] with limited derivations. The linear and angular momentum equations are still given by Eqs. (1.6.14) and (1.6.15), as follows:

$$T_{ji,j} + f_i = \rho \ddot{u}_i, \qquad (2.4.1)$$

$$T_{ij} = T_{ji}. \qquad (2.4.2)$$

The first and second laws of thermodynamics take the following form:

$$\frac{\partial U}{\partial t} = T_{ij} v_{j,i} + r - h_{i,i}, \qquad (2.4.3)$$

$$\dot{\eta} \geq \frac{r}{T} - \left(\frac{h_i}{T}\right)_{,i}. \qquad (2.4.4)$$

The elimination of r from Eqs. (2.4.3) and (2.4.4) gives the Clausius–Duhem inequality:

$$T\dot{\eta} - \dot{U} + T_{ij} v_{j,i} - \frac{h_i T_{,i}}{T} \geq 0. \qquad (2.4.5)$$

With the Helmholtz free energy

$$F(\mathbf{S}, T) = U(\mathbf{S}, \eta) - T\eta, \qquad (2.4.6)$$

Eqs. (2.4.3) and (2.4.5) become

$$\dot{F} + \eta \dot{T} + \dot{\eta} T = T_{ij} v_{j,i} + r - h_{i,i}, \qquad (2.4.7)$$

$$-(\dot{F} + \eta \dot{T}) + T_{ij} v_{j,i} - \frac{h_i T_{,i}}{T} \geq 0. \qquad (2.4.8)$$

With

$$F = F(\mathbf{S}, \ T), \qquad (2.4.9)$$

Eq. (2.4.8) can be written as

$$-\left(\eta + \frac{\partial F}{\partial T}\right)\dot{T} + \left(T_{kl} - \frac{\partial F}{\partial S_{kl}}\right)\dot{S}_{kl} - \frac{1}{T}h_k T_{,k} \geq 0. \qquad (2.4.10)$$

For the inequality to hold for any $\dot{T}$ and any $\dot{S}_{kl}$,

$$\eta = -\frac{\partial F}{\partial T}, \quad T_{kl} = \frac{\partial F}{\partial S_{kl}}. \qquad (2.4.11)$$

Then Eq. (2.4.10) reduces to

$$-\frac{1}{T}h_k T_{,k} \geq 0. \tag{2.4.12}$$

Under Eq. (2.4.11), the energy equation in Eq. (2.4.7) reduces to

$$T\dot{\eta} = r - h_{k,k}. \tag{2.4.13}$$

In order to linearize the constitutive relations, we expand F into a power series about $T = \Theta_0$ and $\mathbf{S} = 0$, where Θ_0 is a uniform reference temperature. Denoting $\theta = T - \Theta_0$, for small θ, keeping quadratic terms only, we can write

$$F = \frac{1}{2}c_{ijkl}S_{ij}S_{kl} - \lambda_{kl}S_{kl}\theta - \frac{1}{2}\frac{C_V}{\Theta_0}\theta^2, \tag{2.4.14}$$

where λ_{ij} are the thermoelastic constants. Equations (2.4.14) and (2.4.11) produce

$$T_{ij} = c_{ijkl}S_{kl} - \lambda_{ij}\theta,$$
$$\eta = \frac{C_V}{\Theta_0}\theta + \lambda_{kl}S_{kl}. \tag{2.4.15}$$

The three-dimensional Fourier's law is

$$h_k = -\kappa_{kl}\theta_{,l}, \tag{2.4.16}$$

which is restricted by Eq. (2.4.12) as follows:

$$-\frac{1}{T}h_k T_{,k} = \frac{1}{T}\kappa_{kl}\theta_{,l}\theta_{,k} \geq 0. \tag{2.4.17}$$

Equation (2.4.17) shows that only the symmetric part of κ_{kl} matters and that κ_{kl} is positive definite.

In summary, for a linear theory, the equations of thermoelasticity are

$$T_{ji,j} + f_i = \rho\ddot{u}_i, \tag{2.4.18}$$
$$\Theta_0\dot{\eta} = r - h_{k,k},$$

$$T_{ij} = c_{ijkl}S_{kl} - \lambda_{ij}\theta,$$
$$\eta = \frac{C_V}{\Theta_0}\theta + \lambda_{kl}S_{kl}, \tag{2.4.19}$$

$$h_k = -\kappa_{kl}\theta_{,l}, \tag{2.4.20}$$
$$S_{ij} = (u_{j,i} + u_{i,j})/2. \tag{2.4.21}$$

With successive substitutions from Eqs. (2.4.19)–(2.4.21), we can write Eq. (2.4.18) as four equations for u_k and θ.

In the compact matrix notation, Eq. $(2.4.19)_1$ can be written as

$$
\begin{bmatrix} T_1 \\ T_2 \\ T_3 \\ T_4 \\ T_5 \\ T_6 \end{bmatrix} = \begin{bmatrix} c_{11} & c_{12} & c_{13} & c_{14} & c_{15} & c_{16} \\ c_{21} & c_{22} & c_{23} & c_{24} & c_{25} & c_{26} \\ c_{31} & c_{32} & c_{33} & c_{34} & c_{35} & c_{36} \\ c_{41} & c_{42} & c_{43} & c_{44} & c_{45} & c_{46} \\ c_{51} & c_{52} & c_{53} & c_{54} & c_{55} & c_{56} \\ c_{61} & c_{62} & c_{63} & c_{64} & c_{65} & c_{66} \end{bmatrix} \begin{bmatrix} S_1 \\ S_2 \\ S_3 \\ S_4 \\ S_5 \\ S_6 \end{bmatrix} - \begin{bmatrix} \lambda_1 \\ \lambda_2 \\ \lambda_3 \\ \lambda_4 \\ \lambda_5 \\ \lambda_6 \end{bmatrix} \theta. \qquad (2.4.22)
$$

Its inverse can be written as

$$
\begin{bmatrix} S_1 \\ S_2 \\ S_3 \\ S_4 \\ S_5 \\ S_6 \end{bmatrix} = \begin{bmatrix} s_{11} & s_{12} & s_{13} & s_{14} & s_{15} & s_{16} \\ s_{21} & s_{22} & s_{23} & s_{24} & s_{25} & s_{26} \\ s_{31} & s_{32} & s_{33} & s_{34} & s_{35} & s_{36} \\ s_{41} & s_{42} & s_{43} & s_{44} & s_{45} & s_{46} \\ s_{51} & s_{52} & s_{53} & s_{54} & s_{55} & s_{56} \\ s_{61} & s_{62} & s_{63} & s_{64} & s_{65} & s_{66} \end{bmatrix} \begin{bmatrix} T_1 \\ T_2 \\ T_3 \\ T_4 \\ T_5 \\ T_6 \end{bmatrix} + \begin{bmatrix} \alpha_1 \\ \alpha_2 \\ \alpha_3 \\ \alpha_4 \\ \alpha_5 \\ \alpha_6 \end{bmatrix} \theta, \qquad (2.4.23)
$$

where α_p are the coefficients of thermal expansion.

2.5 Extension and Bending of Thermoelastic Beams

In this section, we study the behavior of the beam in Fig. 2.4 under a nonuniform temperature change varying linearly along the thickness coordinate x_3. The beam may be in both extension and bending. We derive a set of one-dimensional equations from the three-dimensional theory.

The temperature field is assumed known and is given by

$$\theta = \theta^{(0)}(x_1, t) + x_3 \theta^{(1)}(x_1, t). \qquad (2.5.1)$$

For extension and bending with shear deformation in the (x_1, x_3) plane, the axial and flexural displacements are approximated by

$$u_1 \cong u(x_1, t) + x_3 \psi(x_1, t),$$
$$u_3 \cong w(x_1, t). \qquad (2.5.2)$$

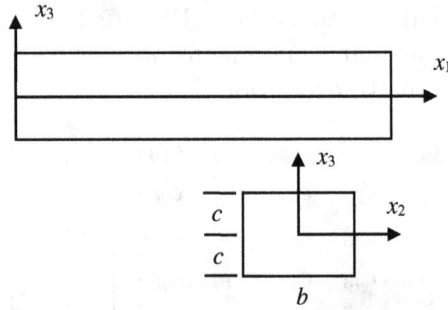

Fig. 2.4.   Side view and cross-section of a rectangular beam.

The axial strain and the shear strain associated with bending are given by

$$S_1 = u_{1,1} = u_{,1} + x_3\psi_{,1},$$
$$S_5 = 2S_{13} = u_{1,3} + u_{3,1} = \psi + w_{,1}. \tag{2.5.3}$$

We limit ourselves to materials with relatively high crystal symmetry. They do not exhibit couplings between extension and shear, and couplings among shears. We also limit ourselves to the case that a temperature change does not induce shear deformation. The constitutive relation is described by

$$
\begin{bmatrix} S_1 \\ S_2 \\ S_3 \\ S_4 \\ S_5 \\ S_6 \end{bmatrix} =
\begin{bmatrix}
s_{11} & s_{12} & s_{13} & 0 & 0 & 0 \\
s_{21} & s_{22} & s_{23} & 0 & 0 & 0 \\
s_{31} & s_{32} & s_{33} & 0 & 0 & 0 \\
0 & 0 & 0 & s_{44} & 0 & 0 \\
0 & 0 & 0 & 0 & s_{55} & 0 \\
0 & 0 & 0 & 0 & 0 & s_{66}
\end{bmatrix}
\begin{bmatrix} T_1 \\ T_2 \\ T_3 \\ T_4 \\ T_5 \\ T_6 \end{bmatrix} +
\begin{bmatrix} \alpha_1 \\ \alpha_2 \\ \alpha_3 \\ 0 \\ 0 \\ 0 \end{bmatrix} \theta. \tag{2.5.4}
$$

For thin beams, we assume

$$T_2 = T_3 = 0. \tag{2.5.5}$$

Then, from Eq. (2.5.4),

$$S_1 = s_{11}T_1 + \alpha_1\theta,$$
$$S_5 = s_{55}T_5. \tag{2.5.6}$$

We solve Eq. (2.5.6) for

$$T_1 = \bar{c}_{11}S_1 - \bar{\lambda}_1\theta,$$
$$T_5 = \bar{c}_{55}S_5, \tag{2.5.7}$$

where

$$\bar{c}_{11} = \frac{1}{s_{11}}, \quad \bar{c}_{55} = \frac{1}{s_{55}}, \quad \bar{\lambda}_1 = \frac{\alpha_1}{s_{11}}. \tag{2.5.8}$$

Substituting from Eqs. (2.5.1) and (2.5.3), we obtain

$$T_1 = \bar{c}_{11}(u_{,1} + x_3\psi_{,1}) - \bar{\lambda}_1(\theta^{(0)} + x_3\theta^{(1)})$$
$$= \bar{c}_{11}u_{,1} - \bar{\lambda}_1\theta^{(0)} + x_3(\bar{c}_{11}\psi_{,1} - \bar{\lambda}_1\theta^{(1)}), \tag{2.5.9}$$
$$T_5 = \bar{c}_{55}(\psi + w_{,1}).$$

The axial force N, transverse shear force Q and bending moment M are defined by the following integrations over a cross-section:

$$N = \int_A T_{11}\,dA = A(\bar{c}_{11}u_{,1} - \bar{\lambda}_1\theta^{(0)}), \tag{2.5.10}$$

$$Q = \int_A T_{13}\,dA = A\bar{c}_{55}(\psi + w_{,1}), \tag{2.5.11}$$

$$M = \int_A x_3 T_1\,dA = \int_{-c}^{c} x_3^2(\bar{c}_{11}\psi_{,1} - \bar{\lambda}_1\theta^{(1)})b\,dx_3$$
$$= \frac{2bc^3}{3}(\bar{c}_{11}\psi_{,1} - \bar{\lambda}_1\theta^{(1)}) = I(\bar{c}_{11}\psi_{,1} - \bar{\lambda}_1\theta^{(1)}), \tag{2.5.12}$$

where

$$A = 2bc, \quad I = \frac{2bc^3}{3}. \tag{2.5.13}$$

From the equations of motion (Newton's second law) of the differential element of the beam in Fig. 2.5 in the x_1 and x_3 directions as well as its moment equation about the x_2 axis, we obtain

$$N_{,1} + F_1 = \rho A\ddot{u},$$
$$Q_{,1} + F_3 = \rho A\ddot{w}, \tag{2.5.14}$$
$$M_{,1} - Q + m_2 = \rho I\ddot{\omega}_2 = \rho I\ddot{\psi},$$

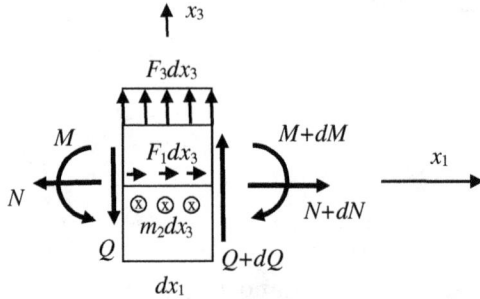

Fig. 2.5. A differential element of the beam under mechanical loads.

where $F_1(x_1,t)$ and $F_3(x_1,t)$ are the axial and transverse loads per unit length of the beam. $m_2(x_1,t)$ is the distributed moment per unit length of the beam.

For materials described by Eq. (2.5.4), the above equations split into two groups. One is for extension under $\theta^{(0)}$, the other is for bending under $\theta^{(1)}$.

The equations for extension are

$$N_{,1} + F_1 = \rho A \ddot{u}, \tag{2.5.15}$$

$$N = A(\bar{c}_{11} u_{,1} - \bar{\lambda}_1 \theta^{(0)}). \tag{2.5.16}$$

The substitution of Eq. (2.5.16) into Eq. (2.5.15) yields a single equation for u, as follows:

$$A(\bar{c}_{11} u_{,11} - \bar{\lambda}_1 \theta^{(0)}_{,1}) + F_1 = \rho A \ddot{u}. \tag{2.5.17}$$

The equations for bending are

$$Q_{,1} + F_3 = \rho A \ddot{w}, \tag{2.5.18}$$

$$M_{,1} - Q + m_2 = \rho I \ddot{\psi},$$

$$Q = A\bar{c}_{55}(\psi + w_{,1}),$$
$$M = I(\bar{c}_{11} \psi_{,1} - \bar{\lambda}_1 \theta^{(1)}), \tag{2.5.19}$$

which can be written as two equations for w and ψ.

For very thin beams, the shear strain can be set to zero

$$S_5 = \psi + w_{,1} = 0, \qquad (2.5.20)$$

which leads to

$$\psi = -w_{,1}, \qquad (2.5.21)$$

$$M = I(\bar{c}_{11}\psi_{,1} - \bar{\lambda}_1\theta^{(1)}) = I(-\bar{c}_{11}w_{,11} - \bar{\lambda}_1\theta^{(1)}). \qquad (2.5.22)$$

For thin beams, we also set the rotatory inertia I in Eq. $(2.5.18)_2$ to zero, which implies the following shear force-bending moment relation or the constitutive relation for Q:

$$Q = M_{,1} + m_2 = I(-\bar{c}_{11}w_{,111} - \bar{\lambda}_1\theta^{(1)}_{,1}) + m_2. \qquad (2.5.23)$$

The substitution of Eq. (2.5.23) into Eq. $(2.5.18)_1$ gives a single equation for w, as follows:

$$I(-\bar{c}_{11}w_{,1111} - \bar{\lambda}_1\theta^{(1)}_{,11}) + m_{2,1} + F_3 = \rho A\ddot{w}. \qquad (2.5.24)$$

2.6 Temperature-Induced Extension of a Composite Rod

In this section, we study the effects of a known and uniform temperature change θ on the extensional deformation of the composite rod shown in Fig. 2.6 [11,12]. It consists of one layer between two identical layers of another material. The material constants of the outer layers will carry a superscript 1 in parentheses. Those of the middle layer will carry a superscript 2 in parentheses.

For extension, we have, approximately,

$$u_1 \cong u(x,t). \qquad (2.6.1)$$

Then the axial strain for the entire rod is given by

$$S = S_1 = u_{1,1} = u_{,x}. \qquad (2.6.2)$$

For the outer layers, in terms of the elastic compliance, the constitutive relation for the axial strain is

$$S = s^{(1)}T + \alpha^{(1)}\theta, \qquad (2.6.3)$$

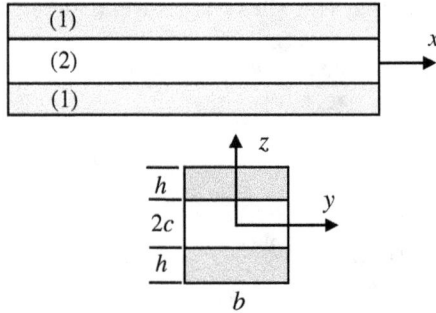

Fig. 2.6.   Side view and cross-section of a composite rod.

where $T = T_1$ is the axial stress. T_2 and T_3 have been set to zero. From Eq. (2.6.3), we solve for

$$T = \bar{c}^{(1)}S - \bar{\lambda}^{(1)}\theta, \tag{2.6.4}$$

where

$$\bar{c}^{(1)} = \frac{1}{s^{(1)}}, \quad \bar{\lambda}^{(1)} = \frac{\alpha^{(1)}}{s^{(1)}}. \tag{2.6.5}$$

Similarly, for the middle layer, the constitutive relation for the axial strain is

$$S = s^{(2)}T + \alpha^{(2)}\theta. \tag{2.6.6}$$

Equation (2.6.6) can be rewritten as

$$T = \bar{c}^{(2)}S - \bar{\lambda}^{(2)}\theta, \tag{2.6.7}$$

where

$$\bar{c}^{(2)} = \frac{1}{s^{(2)}}, \quad \bar{\lambda}^{(2)} = \frac{\alpha^{(2)}}{s^{(2)}}. \tag{2.6.8}$$

The total axial force N in the composite rod are defined by the integration of T over the entire cross-section, and is found to be

$$N = \hat{c}S - \hat{\lambda}\theta, \tag{2.6.9}$$

where

$$\hat{c} = \bar{c}^{(1)}A^{(1)} + \bar{c}^{(2)}A^{(2)},$$
$$\hat{\lambda} = \bar{\lambda}^{(1)}A^{(1)} + \bar{\lambda}^{(2)}A^{(2)}. \tag{2.6.10}$$

$A^{(1)}$ and $A^{(2)}$ are the cross-sectional areas of the outer and middle layers, respectively. The one-dimensional equation of motion is obtained

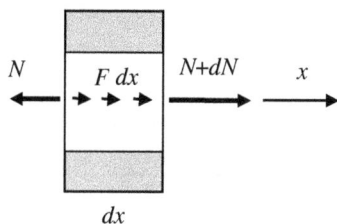

Fig. 2.7. A differential element of the rod with mechanical loads.

by applying Newton's second law to the differential element of the rod shown in Fig. 2.7:

$$\frac{\partial N}{\partial x} + F = 2b \left(\rho^{(1)} h + \rho^{(2)} c \right) \ddot{u}, \qquad (2.6.11)$$

where $F(x,t)$ is the axial load per unit length of the rod. With substitutions from Eqs. (2.6.9) and (2.6.2), we can write Eq. (2.6.11) as an equation for u.

As an example, consider the static extensional deformation of a mechanically free rod within $-L < x < L$. The boundary-value problem is

$$N_{,x} = 0, \quad N = \hat{c}S - \hat{\lambda}\theta, \quad S = u_{,x}, \quad |x| < L,$$
$$N(\pm L) = 0. \qquad (2.6.12)$$

To determine the mechanical displacement uniquely, we set

$$u(0) = 0. \qquad (2.6.13)$$

The solution is simply

$$u = \frac{\hat{\lambda}\theta}{\hat{c}} x. \qquad (2.6.14)$$

Since the layers of a composite rod have different thermal expansion coefficients in general, they are held together by interface shear stresses during thermal expansion which we determine in what follows. As a preparation for determining the interface shear stress, the extensional stress T in the top layer needs to be determined first.

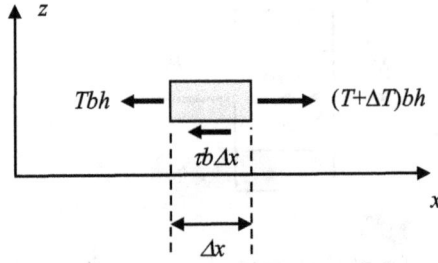

Fig. 2.8.   A differential element of the upper layer at x.

From the constitutive relation of the top layer in Eq. (2.6.4), with the use of Eq. (2.6.14), we obtain

$$T = \bar{c}^{(1)}S - \bar{\lambda}^{(1)}\theta = \bar{c}^{(1)}u_{,x} - \bar{\lambda}^{(1)}\theta$$

$$= \bar{c}^{(1)}\frac{\hat{\lambda}\theta}{\hat{c}} - \bar{\lambda}^{(1)}\theta, \qquad (2.6.15)$$

which is a constant. Then the interface shear stress can be determined as follows. Consider the differential element of the upper layer in Fig. 2.8 in which the distributed interface shear stress is denoted by τ. The equilibrium of the element in the x direction leads to

$$(T + \Delta T)bh - Tbh - \tau b\Delta x = 0, \qquad (2.6.16)$$

which can be rearranged into

$$\tau = \frac{d}{dx}(Th) = 0, \qquad (2.6.17)$$

where Eq. (2.6.15) has been used. Thus, there is no distributed shear stress under the top layer.

When differentiating Eq. (2.6.15) to obtain the zero-shear stress in Eq. (2.6.17), special attention has to be paid to the ends of the top layer where T has a finite discontinuity when it suddenly goes from a finite value in Eq. (2.6.15) to zero as dictated by the free ends. A direct differentiation of Eq. (2.6.15) at the ends produces a singular distribution of τ described by the Dirac delta function which represents a concentrated shear force. Equivalently, this can be treated in a way similar to Fig. 2.8, but the differential element is taken at, e.g., the right end of the top layer as shown in Fig. 2.9. Since the right

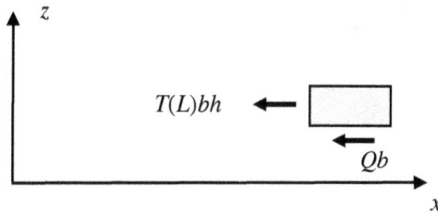

Fig. 2.9.   A differential element of the upper layer at the right end.

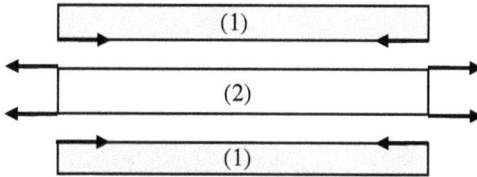

Fig. 2.10.   Interface shear stresses.

end surface is free, for the equilibrium of the element, a concentrated shear force Q per unit dimension along y has to be included. Then the equilibrium equation of the element in the x direction implies that

$$Qb = -T(L)bh = \theta bh \left(\bar{\lambda}^{(1)} - \frac{\bar{c}^{(1)}}{\hat{c}} \hat{\lambda} \right). \tag{2.6.18}$$

Therefore, the layers are held together by end shear forces as shown in Fig. 2.10. We note that this conclusion is based on the one-dimensional model used.

Chapter 3

Electricity and Magnetism

This chapter presents the basics of classical electricity and magnetism in rigid materials without deformation [13–17]. Similar to previous chapters, the present chapter discusses one-dimensional models first and then progresses to three-dimensional theories in general. Fields in free space and basic behaviors of polarization, conduction, semiconduction as well as magnetization in matter are treated. Uncoupled electrostatics and magnetostatics are followed by magnetoelectric, thermal and dynamics couplings.

3.1 Electrostatic Fields in Vacuum

According to Coulomb's law, the electric field at a position $\mathbf{r}$ from a point charge Q (see Fig. 3.1) is given by

$$\mathbf{E} = \frac{Q}{4\pi\varepsilon_0} \frac{\mathbf{r}}{r^3} = \frac{Q}{4\pi\varepsilon_0} \nabla\left(\frac{-1}{r}\right), \qquad (3.1.1)$$

where ε_0 is the electric permittivity of free space, and we have used

$$\frac{\mathbf{r}}{r^3} = \nabla\left(\frac{-1}{r}\right). \qquad (3.1.2)$$

Fig. 3.1.   A point charge Q.

From Eq. (3.1.1), for a closed surface S enclosing Q,

$$\int_S \mathbf{E} \cdot d\mathbf{S} = \int_S \frac{Q}{4\pi\varepsilon_0 r^2} \frac{\mathbf{r}}{r} \cdot d\mathbf{S}$$

$$= \int_S \frac{Q}{4\pi\varepsilon_0} d\Omega = \frac{Q}{4\pi\varepsilon_0} \int_S d\Omega = \frac{Q}{4\pi\varepsilon_0} 4\pi = \frac{Q}{\varepsilon_0},$$

(3.1.3)

where

$$\frac{1}{r^2} \frac{\mathbf{r}}{r} \cdot d\mathbf{S} = d\Omega,$$

(3.1.4)

has been used. $d\Omega$ is the solid angle corresponding to $d\mathbf{S}$. For a closed surface enclosing Q, the solid angle is 4π. The divergence of Eq. (3.1.1) is given by

$$\nabla \cdot \mathbf{E} = \frac{Q}{\varepsilon_0} \nabla \cdot \left(\frac{\mathbf{r}}{4\pi r^3} \right) = \frac{Q}{\varepsilon_0} \delta(\mathbf{r}),$$

(3.1.5)

where

$$\delta(\mathbf{r}) = \frac{1}{4\pi} \nabla \cdot \left(\frac{\mathbf{r}}{r^3} \right) = \frac{1}{4\pi} \nabla \cdot \nabla \left(\frac{-1}{r} \right) = \frac{1}{4\pi} \nabla^2 \left(\frac{-1}{r} \right).$$

(3.1.6)

δ is the Dirac delta function. Mathematically, Eq. (3.1.6) shows that $-1/(4\pi r)$ is the so-called fundamental solution of the Laplace operator ∇^2 (Laplacian). By superposition, in the case of a continuous distribution of charges with density ρ^t occupying a region V, the following are true:

$$\mathbf{E}(\mathbf{x}) = \int_V \frac{\rho^t(\mathbf{x}')}{4\pi\varepsilon_0 r^2} \frac{\mathbf{r}}{r} dV',$$

(3.1.7)

$$\nabla \cdot \mathbf{E} = \frac{\rho^t}{\varepsilon_0},$$

(3.1.8)

where $\mathbf{r} = \mathbf{x} - \mathbf{x}'$ as shown in Fig. 3.2, and the operator ∇ is with respect to $\mathbf{x}$.

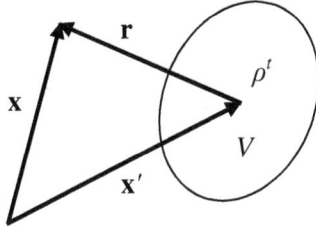

Fig. 3.2.   A continuous distribution of charges in a region V.

From Eq. (3.1.1), for a closed curve C, we have

$$
\begin{aligned}
\oint_C \mathbf{E} \cdot \mathbf{dl} &= \oint_C \frac{Q}{4\pi\varepsilon_0 r^2} \frac{\mathbf{r}}{r} \cdot \mathbf{dl} \\
&= \oint_C \frac{Q}{4\pi\varepsilon_0} \frac{1}{r^2} \, dr = -\oint_C \frac{Q}{4\pi\varepsilon_0} d\left(\frac{1}{r}\right) = 0,
\end{aligned}
\tag{3.1.9}
$$

where $\mathbf{dl}$ is the differential line element along the curve and

$$
\frac{\mathbf{r}}{r} \cdot \mathbf{dl} = dr,
\tag{3.1.10}
$$

has been used. Equation (3.1.9) implies, through Stokes's theorem in calculus,

$$
\nabla \times \mathbf{E} = 0.
\tag{3.1.11}
$$

Then an electrostatic potential φ can be introduced such that

$$
\mathbf{E} = -\nabla\varphi.
\tag{3.1.12}
$$

The substitution of Eq. (3.1.12) into Eq. (3.1.8) results in a single equation for φ as follows:

$$
\nabla \cdot \mathbf{E} = -\nabla \cdot (\nabla\varphi) = -\nabla^2\varphi = \frac{\rho^t}{\varepsilon_0}.
\tag{3.1.13}
$$

3.2   One-Dimensional Dielectrics

When a dielectric is placed in an electric field, the electric charges in its molecules redistribute themselves microscopically, resulting in a

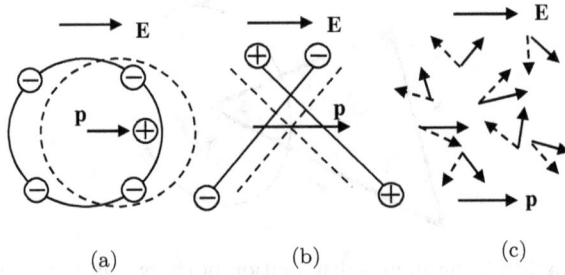

Fig. 3.3.   Microscopic polarizations. (a) Electronic. (b) Ionic. (c) Orientational.

macroscopically polarized state. The microscopic charge redistribution occurs in different ways (see Fig. 3.3).

At the macroscopic level, the distinctions among different polarization mechanisms do not matter. A macroscopic polarization vector per unit volume,

$$\mathbf{P} = \lim_{\Delta V \to 0} \frac{1}{\Delta V} \sum_{\Delta V} \mathbf{p}, \qquad (3.2.1)$$

is introduced which describes the macroscopic polarizing state of the material.

The effect of the polarization vector $\mathbf{P}$ can be represented by the so-called effective volume polarization charge density ρ^P and surface polarization charge density σ^P. We calculate these effective charges using the following one-dimensional model. Consider the differential element of a rod with a cross-section A shown in Fig. 3.4. Let the number of molecules per unit volume be n. Each molecule has equal and opposite charges $\pm q$ which can displace slightly through a distance l relative to each other in the axial direction. Then the axial polarization is $P = qln$. For simplicity, let the negative charges be fixed. At the left end the positive charges within a distance l move into the element according to $PA = qlnA$ there. Similarly, at the right end the positive charges move out of the element according to $(P + dP)A = q(l + dl)nA$. Hence,

$$\rho^P = \frac{qlnA - q(l + dl)nA}{A\,dx} = \frac{PA - (P + dP)A}{A\,dx} = -\frac{dP}{dx}. \qquad (3.2.2)$$

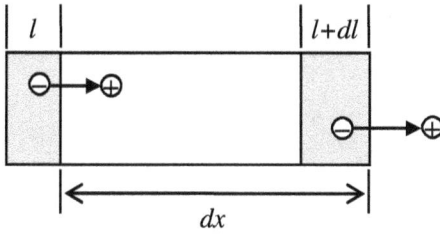

Fig. 3.4. Effective volume polarization charge density ρ^P.

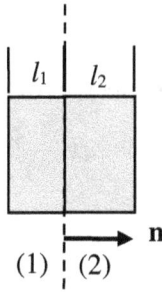

Fig. 3.5. Effective surface polarization charge density σ^P at a discontinuity surface.

Equation (3.2.2) is for a smoothly varying P. The case when P has a discontinuity at an interface between medium (1) and medium (2) is shown in Fig. 3.5. From left to right, across the discontinuity surface with an area A, the axial polarization jumps from P_1 to P_2. The effective polarization charges accumulated near the interface are

$$ql_1 nA - ql_2 nA = P_1 A - P_2 A = A(\mathbf{P}_1 - \mathbf{P}_2) \cdot \mathbf{n} = A\sigma^P, \quad (3.2.3)$$

where

$$\sigma^P = -\mathbf{n} \cdot (\mathbf{P}_2 - \mathbf{P}_1). \quad (3.2.4)$$

In the special case when medium (2) is a vacuum where $\mathbf{P}_2 = 0$, Eq. (3.2.4) reduces to

$$\sigma^P = \mathbf{n} \cdot \mathbf{P}_1. \quad (3.2.5)$$

In a dielectric with effective polarization charges, from Eqs. (3.1.8) and (3.2.2),

$$\frac{dE}{dx} = \frac{\rho^t}{\varepsilon_0} = \frac{1}{\varepsilon_0}(\rho^P + \rho^e) = \frac{1}{\varepsilon_0}\left(-\frac{dP}{dx} + \rho^e\right), \tag{3.2.6}$$

where ρ^e represents charges from origins other than polarization. For dielectrics, ρ^e is zero in most cases, but it is still kept in Eq. (3.2.6) formally. Equation (3.2.6) can be written as

$$\frac{d}{dx}(\varepsilon_0 E + P) = \rho^e. \tag{3.2.7}$$

With the introduction of the electric displacement D by

$$D = \varepsilon_0 E + P, \tag{3.2.8}$$

Eq. (3.2.7) becomes

$$\frac{dD}{dx} = \rho^e. \tag{3.2.9}$$

For a linear material,

$$P = \varepsilon_0 \chi^e E, \tag{3.2.10}$$

where χ^e is the electric susceptibility. Then

$$D = \varepsilon_0(1 + \chi^e)E = \varepsilon E, \tag{3.2.11}$$

where

$$\varepsilon = \varepsilon_0(1 + \chi^e), \tag{3.2.12}$$

is the dielectric constant.

In summary, the governing equations for a one-dimensional dielectric material are

$$\frac{dD}{dx} = \rho^e,$$
$$D = \varepsilon E, \tag{3.2.13}$$
$$E = -\frac{d\varphi}{dx}.$$

With successive substitutions, a single equation for φ can be obtained from Eq. (3.2.13) as

$$-\frac{d}{dx}\left(\varepsilon\frac{d\varphi}{dx}\right) = \rho^e. \tag{3.2.14}$$

3.3 Three-Dimensional Dielectrics

For a three-dimensional dielectric body occupying a region V with a boundary surface S, the effective polarization charge densities are

$$\rho^p = -P_{i,i} = -\nabla \cdot \mathbf{P}, \quad \sigma^p = n_i P_i = \mathbf{n} \cdot \mathbf{P}, \tag{3.3.1}$$

where $\mathbf{n}$ is the unit outward normal of S. The total effective polarization charge is given by

$$\int_V \rho^p \, dV + \int_S \sigma^p \, dS$$

$$= \int_V (-P_{i,i}) dV + \int_S n_i P_i \, dS = \int_V [(-P_{i,i}) + P_{i,i}] dV = 0, \tag{3.3.2}$$

which is as expected. The total electric moment formed by the effective polarization charges is

$$\int_V x_j \rho^p \, dV + \int_S x_j \sigma^p \, dS$$

$$= \int_V x_j(-P_{i,i}) dV + \int_S x_j n_i P_i \, dS$$

$$= \int_V [x_j(-P_{i,i}) + (x_j P_i)_{,i}] dV \tag{3.3.3}$$

$$= \int_V [x_j(-P_{i,i}) + \delta_{ij} P_i + x_j P_{i,i}] dV = \int_V P_j \, dV,$$

which is also as expected.

The charge equation of electrostatics is

$$\nabla \cdot \mathbf{E} = E_{k,k} = \frac{\rho^t}{\varepsilon_0} = \frac{1}{\varepsilon_0}(\rho^P + \rho^e) = \frac{1}{\varepsilon_0}(-P_{k,k} + \rho^e), \qquad (3.3.4)$$

or

$$(\varepsilon_0 E_k + P_k)_{,k} = \rho^e. \qquad (3.3.5)$$

With the introduction of the electric displacement vector

$$D_k = \varepsilon_0 E_k + P_k, \qquad (3.3.6)$$

Eq. (3.3.5) becomes

$$\nabla \cdot \mathbf{D} = D_{k,k} = \rho^e. \qquad (3.3.7)$$

For a linear material,

$$P_k = \varepsilon_0 \chi_{kl}^e E_l. \qquad (3.3.8)$$

Then

$$\begin{aligned} D_k &= \varepsilon_0 E_k + P_k = \varepsilon_0 \delta_{kl} E_l + \varepsilon_0 \chi_{kl}^e E_l \\ &= (\varepsilon_0 \delta_{kl} + \varepsilon_0 \chi_{kl}^e) E_l = \varepsilon_{kl} E_l, \end{aligned} \qquad (3.3.9)$$

where

$$\varepsilon_{kl} = \varepsilon_0 (\delta_{kl} + \chi_{kl}^e). \qquad (3.3.10)$$

In matrix notation,

$$\begin{bmatrix} D_1 \\ D_2 \\ D_3 \end{bmatrix} = \begin{bmatrix} \varepsilon_{11} & \varepsilon_{12} & \varepsilon_{13} \\ \varepsilon_{21} & \varepsilon_{22} & \varepsilon_{23} \\ \varepsilon_{31} & \varepsilon_{32} & \varepsilon_{33} \end{bmatrix} \begin{bmatrix} E_1 \\ E_2 \\ E_3 \end{bmatrix}. \qquad (3.3.11)$$

The energy density of a dielectric is (see Eqs. (3.10.9) and (3.10.10))

$$\hat{U}(\mathbf{D}) = \frac{1}{2} E_i D_i. \qquad (3.3.12)$$

An electric enthalpy density can be introduced through

$$H(\mathbf{E}) = \hat{U} - E_i D_i = -\frac{1}{2} \varepsilon_{ij} E_i E_j, \qquad (3.3.13)$$

which shows that only the symmetric part of ε_{kl} matters. Then

$$D_i = -\frac{\partial H}{\partial E_i} = \varepsilon_{ik} E_k. \tag{3.3.14}$$

The electric field–potential relation is

$$\mathbf{E} = -\nabla\varphi, \quad E_i = -\varphi_{,i}. \tag{3.3.15}$$

Then

$$D_k = \varepsilon_{kl} E_l = -\varepsilon_{kl}\varphi_{,l}. \tag{3.3.16}$$

Corresponding to Eq. (3.2.13), we have Eqs. (3.3.7), (3.3.14) and (3.3.15). The substitution of Eq. (3.3.16) into Eq. (3.3.7) leads to a single equation for φ, as follows:

$$-(\varepsilon_{kl}\varphi_{,l})_{,k} = \rho^e. \tag{3.3.17}$$

3.4 One-Dimensional Conductors

In conductors there are ions fixed to the lattice and free electrons that can flow through the lattice. The ions and electrons carry opposite charges. Consider the case when a conductor is electrically neutral in a reference state. Under a voltage or electric field, the electrons move to form currents and charge distributions. In this section, we analyze the one-dimensional case of a wire with a differential element shown in Fig. 3.6. Metals can also have polarization. The electric field–potential relation and the charge equation of electrostatics are the same as those in Sec. 3.2:

$$E = -\frac{d\varphi}{dx}, \tag{3.4.1}$$

$$\frac{dD}{dx} = -\frac{d}{dx}\left(\varepsilon\frac{d\varphi}{dx}\right) = \rho^e. \tag{3.4.2}$$

Let the cross-sectional area of the wire be A and the axial current density be J. The conservation of charge (continuity equation) of the

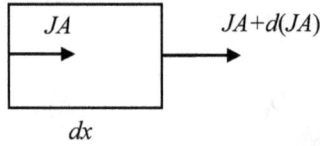

Fig. 3.6.   A differential element of a conducting wire.

differential element in Fig. 3.6 is

$$JA\,dt - [JA + d(JA)]dt = (A\,dx)[\rho^e(t+dt) - \rho^e(t)], \qquad (3.4.3)$$

or

$$-\frac{\partial(JA)}{\partial x}dx\,dt = (A\,dx)\frac{\partial\rho^e}{\partial t}dt, \qquad (3.4.4)$$

$$A\frac{\partial\rho^e}{\partial t} = -\frac{\partial(JA)}{\partial x}. \qquad (3.4.5)$$

The constitutive relation for the current density is Ohm's law:

$$J = \sigma E = -\sigma\frac{d\varphi}{dx}, \qquad (3.4.6)$$

where σ is the conductivity of the material. The substitution of Eq. (3.4.6) into Eq. (3.4.5) yields

$$A\frac{\partial\rho^e}{\partial t} = \frac{\partial}{\partial x}\left(\sigma\frac{\partial\varphi}{\partial x}A\right). \qquad (3.4.7)$$

Equations (3.4.2) and (3.4.7) are two equations for ρ^e and φ.

As an example, consider static conduction in a homogeneous wire under a voltage V. The boundary-value problem of Eqs. (3.4.2) and (3.4.7) is

$$\begin{aligned} -\varepsilon\varphi'' &= \rho^e, \quad 0 < x < L, \\ \sigma A\varphi'' &= 0, \quad 0 < x < L, \\ \varphi(0) &= 0, \quad \varphi(L) = V. \end{aligned} \qquad (3.4.8)$$

The solution is

$$\varphi = \frac{x}{L}V, \quad E = -\frac{V}{L}, \quad J = -\sigma\frac{V}{L}, \quad \rho^e = 0. \qquad (3.4.9)$$

3.5 Three-Dimensional Conductors

For a three-dimensional conductor, Ohm's law is written as

$$J_k = \sigma_{kl} E_l, \tag{3.5.1}$$

or

$$\begin{bmatrix} J_1 \\ J_2 \\ J_3 \end{bmatrix} = \begin{bmatrix} \sigma_{11} & \sigma_{12} & \sigma_{13} \\ \sigma_{21} & \sigma_{22} & \sigma_{23} \\ \sigma_{31} & \sigma_{32} & \sigma_{33} \end{bmatrix} \begin{bmatrix} E_1 \\ E_2 \\ E_3 \end{bmatrix}, \tag{3.5.2}$$

where

$$\sigma_{kl} = \sigma_{lk}. \tag{3.5.3}$$

In terms of the electric potential φ,

$$J_k = \sigma_{kl} E_l = -\sigma_{kl}\varphi_{,l}. \tag{3.5.4}$$

The continuity equation takes the following form:

$$\frac{\partial \rho^e}{\partial t} = -J_{k,k}, \tag{3.5.5}$$

which, with the use of Eq. (3.5.4), becomes

$$\frac{\partial \rho^e}{\partial t} = (\sigma_{kl}\varphi_{,l})_{,k}. \tag{3.5.6}$$

The charge equation of electrostatics is

$$D_{k,k} = -(\varepsilon_{kl}\varphi_{,l})_{,k} = \rho^e. \tag{3.5.7}$$

Equations (3.5.6) and (3.5.7) are two equations for ρ^e and φ.

In the special case of an isotropic conductor, we have

$$\varepsilon_{kl} = \varepsilon\delta_{lk}, \quad \sigma_{kl} = \sigma\delta_{lk}. \tag{3.5.8}$$

Then

$$\varepsilon\nabla \cdot \mathbf{E} = \rho^e,$$
$$\dot{\rho}^e = -\nabla \cdot \mathbf{J}, \tag{3.5.9}$$
$$\mathbf{J} = \sigma\mathbf{E}.$$

With substitutions from Eqs. $(3.5.9)_{1,3}$, we can write Eq. $(3.5.9)_2$ as

$$\dot{\rho}^e = -\frac{\sigma}{\varepsilon}\rho^e. \tag{3.5.10}$$

Equation (3.5.10) can be integrated to produce

$$\rho^e(t) = \rho^e(0)\exp\left(-\frac{\sigma}{\varepsilon}t\right) = \rho^e(0)\exp\left(-\frac{t}{\tau}\right), \tag{3.5.11}$$

where

$$\tau = \frac{\varepsilon}{\sigma}, \tag{3.5.12}$$

is the so-called relaxation time of a conductor which describes the time needed to reach an essentially stationary condition after an initial disturbance.

3.6   One-Dimensional Semiconductors

In semiconductors, in addition to the effective polarization charges, there are charges from doping which are imbedded in the lattice and mobile charge carriers of holes and electrons which are responsible for conduction [18–20]. The one-dimensional equations of electrostatics take the following form:

$$E = -\frac{d\varphi}{dx},$$
$$D = \varepsilon E, \tag{3.6.1}$$
$$\frac{dD}{dx} = \rho^e = q(p - n + N_D^+ - N_A^-),$$

where q is the elementary charge. p and n are the concentrations of holes and electrons. N_A^- and N_D^+ are the concentrations of ionized acceptors and donors from doping. N_A^- and N_D^+ produce holes and electrons which form p and n, respectively. There may be charges of other origins which are not considered here. The conservation of charges for holes and electrons (continuity equations) are, respectively,

$$q\dot{p} = -\frac{dJ^p}{dx} + \gamma^p,$$
$$q\dot{n} = \frac{dJ^n}{dx} + \gamma^n,$$

(3.6.2)

where J^p and J^n are the hole and electron current densities. γ^p and γ^n are the sources of holes and electrons. They may be from mechanical, thermal, electrical, magnetic and optical origins. For a macroscopic theory, γ^p and γ^n belong to the so-called constitutive relations whose expressions are determined experimentally. The constitutive relations for the current densities are

$$J^p = qp\mu^p E - qD^p \frac{dp}{dx},$$
$$J^n = qn\mu^n E + qD^n \frac{dn}{dx},$$

(3.6.3)

where μ^p and μ^n are the mobility of holes and electrons, respectively. The first term in J^p (or J^n) is the drift current which is nonlinear as a product of the carrier concentration p (or n) and the electric field E. The second term in J^p (or J^n) is the diffusion current. D^p and D^n are the diffusion constants. They satisfy the following Einstein relation:

$$\frac{\mu^p}{D^p} = \frac{\mu^n}{D^n} = \frac{q}{k_B T},$$

(3.6.4)

where k_B is the Boltzmann constant and T, the absolute temperature. With substitutions from Eqs. (3.6.1)$_{1,2}$ and (3.6.3), we can write Eqs. (3.6.1)$_3$ and (3.6.3) as three equations for φ, p and n.

In the above theory, N_A^- and N_D^+ appear as electrical loads or nonhomogeneous terms of the equations. As an essentially linear theory (except the drift currents) valid for weak electric fields, in general the loads cannot be large or there cannot be high concentrations of charges. The gradient terms of dp/dx and dn/dx should also be small so that the current densities are small. The rest of this book will be limited to relatively simple situations in which both γ^p and γ^n vanish. For low charge concentrations, it is sufficient to assume that both N_A^- and N_D^+ are small (low doping). Then p and n are also small. Hence, ρ^e in Eq. (3.6.1)$_3$ is small. In this case, it may be argued that as second-order terms the nonlinear drift currents are negligible, which leads to a fully linear theory with diffusion-dominated currents. Such a theory may be too crude for many problems and is not pursued in this book. In the dielectric limit when both N_A^- and N_D^+ are zero, the equations become homogeneous which can be satisfied by vanishing fields of p and n without currents.

A more common situation is that N_A^- and N_D^+ are not small. In this case, the following linearized and hence simpler version of Eq. (3.6.3) may be introduced which is sufficient for certain purposes. For convenience we denote

$$p_0 = N_A^-, \quad n_0 = N_D^+. \tag{3.6.5}$$

We assume that the material is homogeneous and electrically neutral in the reference state with $p = p_0$ and $n = n_0$. In the reference state we also have $E = 0$. Under some electrical disturbance, E develops. At the same time p and n change from p_0 and n_0 to

$$p = p_0 + \Delta p, \quad n = n_0 + \Delta n, \tag{3.6.6}$$

where Δp and Δn are assumed to be small. We linearize Eq. (3.6.3) as

$$
\begin{aligned}
J^p &\cong q p_0 \mu^p E - q D^p \frac{d}{dx}(p_0 + \Delta p), \\
J^n &\cong q n_0 \mu^n E + q D^n \frac{d}{dx}(n_0 + \Delta n).
\end{aligned}
\tag{3.6.7}
$$

Since p_0 and n_0 are assumed to be uniform, Eq. (3.6.7) becomes homogeneous in E, Δp and Δn, as follows:

$$
\begin{aligned}
J^p &\cong q p_0 \mu^p E - q D^p \frac{d}{dx}(\Delta p), \\
J^n &\cong q n_0 \mu^n E + q D^n \frac{d}{dx}(\Delta n).
\end{aligned}
\tag{3.6.8}
$$

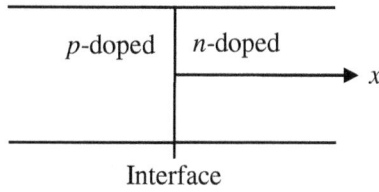

Fig. 3.7.   An interface between two semiconductors.

Equation (3.6.6) also makes the right-hand side of Eq. (3.6.1)$_3$ homogeneous. With Δp and Δn, Eqs. (3.6.1)$_3$ and (3.6.2) become

$$\frac{dD}{dx} = \rho^e = q(\Delta p - \Delta n), \tag{3.6.9}$$

$$q\frac{\partial(\Delta p)}{\partial t} = -\frac{dJ^p}{dx},$$
$$q\frac{\partial(\Delta n)}{\partial t} = \frac{dJ^n}{dx}, \tag{3.6.10}$$

where ρ^e is small because Δp and Δn are small. Equations (3.6.9) and (3.6.10) can be written as three linear equations for φ, Δp and Δn with the use of Eqs. (3.6.1)$_{1,2}$ and (3.6.8).

As an example, consider the interface between two semiconductors in Fig. 3.7 [21]. The two materials are doped into a p region where $p > n$, and an n region where $n > p$. Diffusion occurs at the interface to form a PN junction. The doping in each region is uniform, but it has a jump discontinuity at the interface forming an abrupt junction.

Mathematically, we need to treat the two regions separately and then apply boundary conditions at infinity as well as continuity conditions at the interface. The governing equations for φ, Δp and Δn can be written as

$$-\varepsilon\frac{d^2\varphi}{dx^2} = q(\Delta p - \Delta n),$$
$$0 = p_0\mu^p\frac{d^2\varphi}{dx^2} + D^p\frac{d^2}{dx^2}(\Delta p), \tag{3.6.11}$$
$$0 = -n_0\mu^n\frac{d^2\varphi}{dx^2} + D^n\frac{d^2}{dx^2}(\Delta n),$$

which can be rewritten into

$$\frac{d^2}{dx^2}(\Delta p - \Delta n) = k^2 (\Delta p - \Delta n),$$

$$\frac{d^2 \varphi}{dx^2} = -\frac{q}{k^2 \varepsilon} \frac{d^2}{dx^2}(\Delta p - \Delta n), \qquad (3.6.12)$$

$$\frac{d^2}{dx^2}(\Delta n) = \frac{n_0 \mu^n}{D^n} \frac{d^2 \varphi}{dx^2},$$

where

$$k^2 = \left(\frac{p_0 \mu^p}{D^p} + \frac{n_0 \mu^n}{D^n}\right) \frac{q}{\varepsilon} = \frac{1}{\lambda_D^2}. \qquad (3.6.13)$$

λ_D is called the Debye–Hückel length. At infinity we have the following boundary conditions:

$$D(\pm\infty) = 0, \quad J^p(\pm\infty) = 0, \quad J^n(\pm\infty) = 0. \qquad (3.6.14)$$

At the interface we have the following continuity conditions:

$$\varphi(0^-) = \varphi(0^+), \qquad D(0^-) = D(0^+),$$

$$n(0^-) = n(0^+), \qquad p(0^-) = p(0^+), \qquad (3.6.15)$$

$$J^n(0^-) = J^n(0^+), \qquad J^p(0^-) = J^p(0^+).$$

The electric potential may have an arbitrary constant which does not affect the electric field it produces. To make potential unique, we impose

$$\varphi(-\infty) = 0. \qquad (3.6.16)$$

We also have the following conditions representing the global conservation of holes and electrons:

$$\int_{-\infty}^{0} \Delta p \, dx + \int_{0}^{\infty} \Delta p \, dx = 0,$$

$$\int_{-\infty}^{0} \Delta n \, dx + \int_{0}^{\infty} \Delta n \, dx = 0. \qquad (3.6.17)$$

We use a prime for the material parameters in the p-doped left region and a double prime for those in the n-doped region on the right. Equations (3.6.12) are linear ordinary differential equations with constant coefficients. Its general solution can be obtained in a straightforward and systematic manner. For $x < 0$, we have

$$\Delta p - \Delta n = C_1 \exp(k'x), \tag{3.6.18}$$

$$\varphi = -\frac{q}{(k')^2 \varepsilon'} C_1 \exp(k'x) + C_2 x + C_3, \tag{3.6.19}$$

$$\Delta n = \frac{n_0' \mu'^n}{D'^n}\left[-\frac{q}{(k')^2 \varepsilon'} C_1 \exp(k'x) + C_2 x + C_3\right] \tag{3.6.20}$$
$$+ C_6 x + C_7,$$

where C_1 through C_7 are undetermined constants. Similarly, for $x > 0$, we have

$$\Delta p - \Delta n = C_8 \exp(-k''x), \tag{3.6.21}$$

$$\varphi = -\frac{q}{(k'')^2 \varepsilon''} C_8 \exp(-k''x) + C_9 x + C_{10}, \tag{3.6.22}$$

$$\Delta n = \frac{n_0'' \mu''^n}{D''^n}\left[-\frac{q}{(k'')^2 \varepsilon''} C_8 \exp(-k''x) + C_9 x + C_{10}\right] \tag{3.6.23}$$
$$+ C_{13} x + C_{14},$$

where C_8 through C_{14} are undetermined constants. C_1 through C_{14} are determined from Eqs. (3.6.14)–(3.6.17). Consider the doping profile shown in Fig. 3.8. As a special case, when

$$p_0'' = n_0', \quad n_0'' = p_0', \tag{3.6.24}$$

we have

$$k' = k''. \tag{3.6.25}$$

Qualitatively, the solutions for the charges, electric and potential fields (the so-called built-in fields) near the junction from the above one-dimensional model are shown in Figs. 3.9 through 3.11 along with comparisons to the corresponding solutions from the conventional depletion-layer model of PN junctions with assumed charge distributions. For the electric potential, the two models predict similar results and therefore only one is shown.

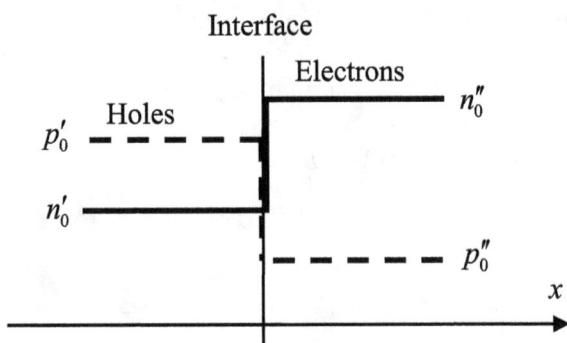

Fig. 3.8.   Doping profile.

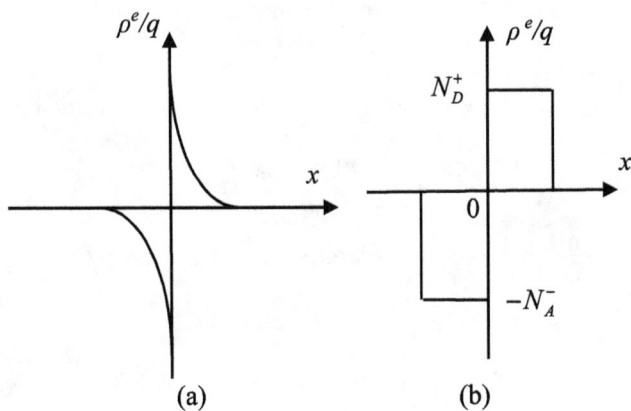

Fig. 3.9.   Charge distribution. (a) One-dimensional model. (b) Depletion-layer model.

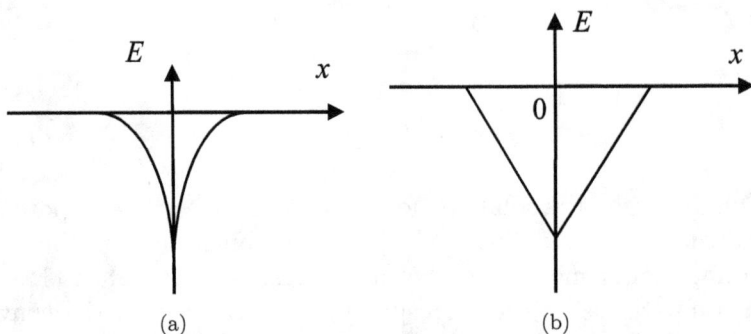

Fig. 3.10.   Electric field. (a) One-dimensional model. (b) Depletion-layer model.

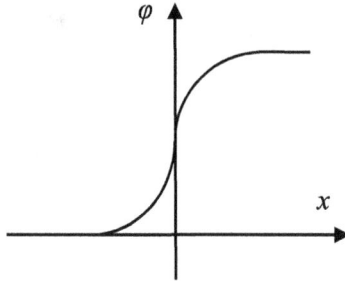

Fig. 3.11. Electric potential.

The above linear analysis of a PN junction is able to show the formation of the built-in charge distributions and electric as well as potential fields in a PN junction. However, to predict the typical nonlinear current–voltage relation of a PN junction, the nonlinear constitutive relations for current densities in Eq. (3.6.3) need to be used, which presents mathematical challenges. The built-in fields in a PN junction may be viewed as pre-existing or initial fields which affect the current when a voltage is applied with the presence of the initial fields. The basic effects of initial fields can be described by the so-called linear theory for small incremental fields superposed on finite initial fields, which is presented as follows. We begin with the following nonlinear formulation:

$$\frac{\partial D}{\partial x} = q(p - n + N_D^+ - N_A^-), \tag{3.6.26}$$

$$D = \varepsilon E, \quad E = -\frac{\partial \varphi}{\partial x}, \tag{3.6.27}$$

$$q\dot{p} = -\frac{\partial J^p}{\partial x}, \quad q\dot{n} = \frac{\partial J^n}{\partial x}, \tag{3.6.28}$$

$$J^p = qp\mu^p E - qD^p \frac{\partial p}{\partial x},$$
$$J^n = qn\mu^n E + qD^n \frac{\partial n}{\partial x}. \tag{3.6.29}$$

Let

$$\varphi(x,t) = \overline{\varphi}(x) + \tilde{\varphi}(x,t),$$
$$E(x,t) = \overline{E}(x) + \tilde{E}(x,t), \tag{3.6.30}$$
$$D(x,t) = \overline{D}(x) + \tilde{D}(x,t),$$

$$p(x, t) = \bar{p}(x) + \tilde{p}(x, t),$$
$$n(x, t) = \bar{n}(x) + \tilde{n}(x, t), \tag{3.6.31}$$

$$J^p(x, t) = \overline{J}^p(x) + \tilde{J}^p(x, t),$$
$$J^n(x, t) = \overline{J}^n(x) + \tilde{J}^n(x, t), \tag{3.6.32}$$

where we have used an over bar to describe the static initial fields, and a tilde for the dynamic incremental fields. Substituting Eqs. (3.6.30)–(3.6.32) into Eqs. (3.6.26)–(3.6.29), collecting zero- and first-order terms of the small incremental fields, we obtain two sets of equations. One is for the static initial fields:

$$\frac{\partial \overline{D}}{\partial x} = q(\bar{p} - \bar{n} + N_D^+ - N_A^-), \tag{3.6.33}$$

$$\overline{D} = \varepsilon \overline{E}, \quad \overline{E} = -\frac{\partial \overline{\varphi}}{\partial x}, \tag{3.6.34}$$

$$0 = -\frac{\partial \overline{J}^p}{\partial x}, \quad 0 = \frac{\partial \overline{J}^n}{\partial x}, \tag{3.6.35}$$

$$\overline{J}^p = q\bar{p}\mu^p \overline{E} - qD^p \frac{\partial \bar{p}}{\partial x},$$
$$\overline{J}^n = q\bar{n}\mu^n \overline{E} + qD^n \frac{\partial \bar{n}}{\partial x}. \tag{3.6.36}$$

The other is for the small incremental fields:

$$\frac{\partial \tilde{D}}{\partial x} = q(\tilde{p} - \tilde{n}), \tag{3.6.37}$$

$$\tilde{D} = \varepsilon \tilde{E}, \quad \tilde{E} = -\frac{\partial \tilde{\varphi}}{\partial x}, \tag{3.6.38}$$

$$q\dot{\tilde{p}} = -\frac{\partial \tilde{J}^p}{\partial x}, \quad q\dot{\tilde{n}} = \frac{\partial \tilde{J}^n}{\partial x}, \tag{3.6.39}$$

$$\tilde{J}^p = q\bar{p}\mu^p \tilde{E} + q\mu^p \overline{E}\tilde{p} - qD^p \frac{\partial \tilde{p}}{\partial x},$$
$$\tilde{J}^n = q\bar{n}\mu^n \tilde{E} + q\mu^n \overline{E}\tilde{n} + qD^n \frac{\partial \tilde{n}}{\partial x}. \tag{3.6.40}$$

Equations (3.6.37)–(3.6.40) are linear in the small incremental fields. Equation (3.6.40) shows that the initial fields appear in the coefficients of the equations for the incremental fields.

3.7 Three-Dimensional Semiconductors

For a three-dimensional body of a semiconductor, the equations of electrostatics are

$$
\begin{aligned}
E_k &= -\varphi_{,k}, \\
D_k &= \varepsilon_{kl}E_l, \\
D_{k,k} &= q(p - n + N_D^+ - N_A^-).
\end{aligned}
\tag{3.7.1}
$$

The continuity equations for holes and electrons are

$$
\begin{aligned}
q\dot{p} &= -J_{i,i}^p + \gamma^p, \\
q\dot{n} &= J_{i,i}^n + \gamma^n.
\end{aligned}
\tag{3.7.2}
$$

The constitutive relations for the current densities are

$$
\begin{aligned}
J_i^p &= qp\mu_{ij}^p E_j - qD_{ij}^p p_{,j}, \\
J_i^n &= qn\mu_{ij}^n E_j + qD_{ij}^n n_{,j}.
\end{aligned}
\tag{3.7.3}
$$

We assume uniform p_0 and n_0 at a reference state with

$$
\begin{aligned}
p &= N_A^- = p_0, \quad n = N_D^+ = n_0, \\
\mathbf{D} &= 0, \quad \mathbf{E} = 0, \quad \mathbf{J}^p = 0, \quad \mathbf{J}^n = 0.
\end{aligned}
\tag{3.7.4}
$$

When small electrical loads are applied, we let

$$
\begin{aligned}
p &= p_0 + \Delta p, \\
n &= n_0 + \Delta n.
\end{aligned}
\tag{3.7.5}
$$

For small Δp and Δn,

$$
\begin{aligned}
J_i^p &\cong qp_0\mu_{ij}^p E_j - qD_{ij}^p(p_0 + \Delta p)_{,j}, \\
J_i^n &\cong qn_0\mu_{ij}^n E_j + qD_{ij}^n(n_0 + \Delta n)_{,j}.
\end{aligned}
\tag{3.7.6}
$$

Since p_0 and n_0 are assumed to be uniform, Eq. (3.7.6) becomes homogeneous in $\mathbf{E}$, Δp and Δn, as follows:

$$J_i^p \cong qp_0\mu_{ij}^p E_j - qD_{ij}^p(\Delta p)_{,j},$$
$$J_i^n \cong qn_0\mu_{ij}^n E_j + qD_{ij}^n(\Delta n)_{,j}. \tag{3.7.7}$$

With Δp and Δn, Eqs. (3.7.1)$_3$ and (3.7.2) become

$$D_{i,i} = q(\Delta p - \Delta n), \tag{3.7.8}$$

$$q\frac{\partial(\Delta p)}{\partial t} = -J_{i,i}^p,$$
$$q\frac{\partial(\Delta n)}{\partial t} = J_{i,i}^n, \tag{3.7.9}$$

which can be written as three linear equations for φ, Δp and Δn with the use of Eqs. (3.7.1)$_{1,2}$ and (3.7.7).

3.8 Magnetostatics

Consider a current density distribution $\mathbf{J}^t$ over a region V. According to the Biot–Savart law, the magnetic induction $\mathbf{B}$ at a point $\mathbf{x}$ is given by

$$\mathbf{B}(\mathbf{x}) = \frac{\mu_0}{4\pi}\int_V \mathbf{J}^t(\mathbf{x}') \times \frac{\mathbf{r}}{r^3}dV'$$
$$= \frac{\mu_0}{4\pi}\int_V \mathbf{J}^t(\mathbf{x}') \times \nabla\left(\frac{-1}{r}\right)dV', \tag{3.8.1}$$

where $\mathbf{r} = \mathbf{x}-\mathbf{x}'$ and Eq. (3.1.2) has been used. The gradient operator ∇ is with respect to $\mathbf{x}$. μ_0 is the magnetic permeability of free space. Then

$$\nabla \cdot \mathbf{B} = \frac{\mu_0}{4\pi}\int_V \nabla \cdot \left\{\mathbf{J}^t(\mathbf{x}') \times \nabla\left(\frac{-1}{r}\right)\right\}dV'$$
$$= \frac{\mu_0}{4\pi}\int_V \left\{[\nabla \times \mathbf{J}^t(\mathbf{x}')] \cdot \frac{\mathbf{r}}{r^3} - \mathbf{J}^t(\mathbf{x}') \cdot \left[\nabla \times \nabla\left(\frac{-1}{r}\right)\right]\right\}dV'$$
$$= \frac{\mu_0}{4\pi}\int_V \{\mathbf{0} - \mathbf{0}\}\,dV' = 0, \tag{3.8.2}$$

where the following vector identity has been used:

$$\nabla \cdot (\mathbf{a} \times \mathbf{b}) = (\nabla \times \mathbf{a}) \cdot \mathbf{b} - \mathbf{a} \cdot (\nabla \times \mathbf{b}). \tag{3.8.3}$$

For any scalar field f and vector field $\mathbf{a}$, we have

$$\nabla \times (f\mathbf{a}) = (\nabla f) \times \mathbf{a} + f(\nabla \times \mathbf{a}),$$
$$\nabla \cdot (f\mathbf{a}) = (\nabla f) \cdot \mathbf{a} + f(\nabla \cdot \mathbf{a}). \tag{3.8.4}$$

According to Eq. $(3.8.4)_1$,

$$\nabla \times \left[\frac{1}{r}\mathbf{J}^t(\mathbf{x}') \right] = \left(\nabla \frac{1}{r} \right) \times \mathbf{J}^t(\mathbf{x}') + \frac{1}{r}\nabla \times \mathbf{J}^t(\mathbf{x}')$$

$$= \left(\nabla \frac{1}{r} \right) \times \mathbf{J}^t(\mathbf{x}') + \mathbf{0} = \left(\nabla \frac{1}{r} \right) \times \mathbf{J}^t(\mathbf{x}'). \tag{3.8.5}$$

Then Eq. (3.8.1) can be written as

$$\mathbf{B}(\mathbf{x}) = \frac{\mu_0}{4\pi} \int_V \nabla \left(\frac{1}{r} \right) \times \mathbf{J}^t(\mathbf{x}')dV' = \frac{\mu_0}{4\pi} \int_V \nabla \times \left[\frac{1}{r}\mathbf{J}^t(\mathbf{x}') \right] dV'$$

$$= \frac{\mu_0}{4\pi} \nabla \times \int_V \frac{\mathbf{J}^t(\mathbf{x}')}{r}dV' = \nabla \times \mathbf{A}, \tag{3.8.6}$$

where we have denoted

$$\mathbf{A} = \frac{\mu_0}{4\pi} \int_V \frac{\mathbf{J}^t(\mathbf{x}')}{r}dV'. \tag{3.8.7}$$

Then the divergence of $\mathbf{B}$ can also be calculated as

$$\nabla \cdot \mathbf{B} = \nabla \cdot (\nabla \times \mathbf{A}) = 0. \tag{3.8.8}$$

By another vector identity, the curl of $\mathbf{B}$ can be written as

$$\nabla \times \mathbf{B} = \nabla \times (\nabla \times \mathbf{A}) = \nabla(\nabla \cdot \mathbf{A}) - \nabla^2 \mathbf{A}. \tag{3.8.9}$$

We have

$$\nabla \cdot \mathbf{A} = \frac{\mu_0}{4\pi} \int_V \nabla \cdot \left[\frac{\mathbf{J}^t(\mathbf{x}')}{r} \right] dV'$$

$$= \frac{\mu_0}{4\pi} \int_V \left[\nabla \left(\frac{1}{r} \right) \cdot \mathbf{J}^t(\mathbf{x}') + \frac{1}{r}\nabla \cdot \mathbf{J}^t(\mathbf{x}') \right] dV'$$

$$= \frac{\mu_0}{4\pi} \int_V \left[-\nabla' \left(\frac{1}{r} \right) \cdot \mathbf{J}^t(\mathbf{x}') + 0 \right] dV'$$

$$= -\frac{\mu_0}{4\pi} \int_V \nabla' \cdot \left[\mathbf{J}^t(\mathbf{x}')\frac{1}{r} \right] dV' + \frac{\mu_0}{4\pi} \int_V \frac{1}{r}\nabla' \cdot \mathbf{J}^t(\mathbf{x}')dV', \tag{3.8.10}$$

where Eq. $(3.8.4)_2$ has been used twice. ∇' is with respect to $\mathbf{x}'$ and

$$\nabla' \frac{1}{r} = -\nabla \frac{1}{r}. \tag{3.8.11}$$

The first term on the right-hand side of Eq. (3.8.10) can be converted to a surface integral. When V includes all currents, there are no currents flowing through the surface of V. Therefore, this term vanishes. The second term on the right-hand side of Eq. (3.8.10) also vanishes because steady-state currents have zero divergence. Therefore,

$$\nabla \cdot \mathbf{A} = 0. \tag{3.8.12}$$

We also have

$$\nabla^2 \mathbf{A} = \frac{\mu_0}{4\pi} \nabla^2 \int_V \frac{\mathbf{J}^t(\mathbf{x}')}{r} dV' = \frac{\mu_0}{4\pi} \int_V \mathbf{J}^t(\mathbf{x}') \nabla^2 \left(\frac{1}{r} \right) dV'$$

$$= \frac{\mu_0}{4\pi} \int_V \mathbf{J}^t(\mathbf{x}') \left[-4\pi \delta(\mathbf{r}) \right] dV' = -\mu_0 \mathbf{J}^t(\mathbf{x}), \tag{3.8.13}$$

where Eq. (3.1.6) has been used. From Eqs. (3.8.9), (3.8.12) and (3.8.13), we have

$$\nabla \times \mathbf{B} = \mu_0 \mathbf{J}^t. \tag{3.8.14}$$

In a dielectric material, at the microscopic level, molecules carry current loops possessing magnetic moments $\mathbf{m}$. For example, using cylindrical coordinates (r, θ, z) with unit vectors $(\mathbf{e}_r, \mathbf{e}_\theta, \mathbf{e}_z)$, the magnetic moment of the circular current loop in Fig. 3.12 is defined by and found to be

$$\mathbf{m} = \frac{1}{2} \oint_C \mathbf{r} \times I dl = \frac{1}{2} \oint_C R \mathbf{e}_r \times I dl \mathbf{e}_\theta$$

$$= \frac{1}{2} RI \mathbf{e}_z \oint_C dl = \frac{1}{2} RI \mathbf{e}_z 2\pi R \tag{3.8.15}$$

$$= I\pi R^2 \mathbf{e}_z = I S \mathbf{e}_z = I\mathbf{S},$$

where $S = \pi R^2$ is the area enclosed by the circle C and $\mathbf{S} = S\mathbf{e}_z$. Equation (3.8.15) in the form of $\mathbf{m} = I\mathbf{S}$ is also valid for current loops of other shapes in the (x, y) plane.

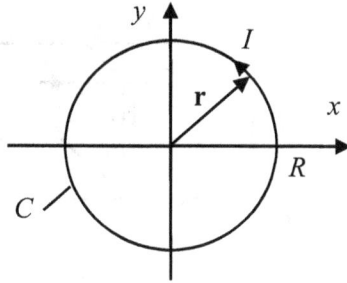

Fig. 3.12. A circular current loop.

In a way similar to what is shown in Fig. 3.3(c), **m** may form a macroscopic magnetization vector **M** defined by

$$\mathbf{M} = \lim_{\Delta V \to 0} \frac{1}{\Delta V} \sum_{\Delta V} \mathbf{m}. \tag{3.8.16}$$

It can be shown that effectively **M** is equivalent to the following magnetization current density:

$$
\begin{aligned}
\mathbf{J}^M &= \nabla \times \mathbf{M} \\
&= \left(\frac{\partial M_z}{\partial y} - \frac{\partial M_y}{\partial z} \right) \mathbf{i} + \left(\frac{\partial M_x}{\partial z} - \frac{\partial M_z}{\partial x} \right) \mathbf{j} + \left(\frac{\partial M_y}{\partial x} - \frac{\partial M_x}{\partial y} \right) \mathbf{k}.
\end{aligned}
\tag{3.8.17}
$$

For example, part of the z (or **k**) component of Eq. (3.8.17) when the variation of M_x along y is considered can be calculated from Fig. 3.13 as follows. We have, from the figure,

$$I' \Delta y\, \Delta z = M_x \Delta x\, \Delta y\, \Delta z, \tag{3.8.18}$$

$$I'' \Delta y\, \Delta z = \left(M_x + \frac{\partial M_x}{\partial y} \Delta y \right) \Delta x\, \Delta y\, \Delta z. \tag{3.8.19}$$

Then

$$I' - I'' = -\frac{\partial M_x}{\partial y} \Delta x\, \Delta y = J_z^M \Delta x\, \Delta y, \tag{3.8.20}$$

or

$$J_z^M = -\frac{\partial M_x}{\partial y}. \tag{3.8.21}$$

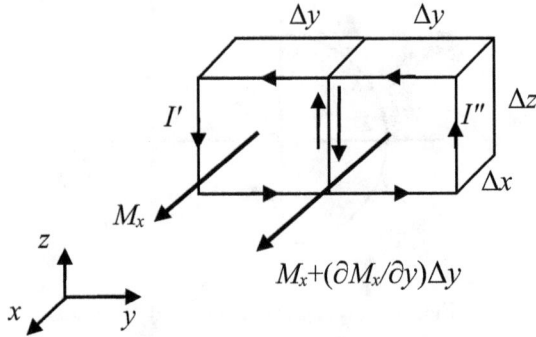

Fig. 3.13. The z component of $\mathbf{J}^M$.

Similarly, when the variation of M_y along x is considered, we have

$$J_z^M = \frac{\partial M_y}{\partial x}. \tag{3.8.22}$$

Adding Eqs. (3.8.21) and (3.8.22), we obtain the z component of Eq. (3.8.17) as follows:

$$J_z^M = \frac{\partial M_y}{\partial x} - \frac{\partial M_x}{\partial y}. \tag{3.8.23}$$

With Eq. (3.8.17), we can write Eq. (3.8.14) as

$$\nabla \times \mathbf{B} = \mu_0(\mathbf{J} + \mathbf{J}^M), \tag{3.8.24}$$

or

$$\nabla \times \left(\frac{1}{\mu_0}\mathbf{B} - \mathbf{M}\right) = \mathbf{J}, \tag{3.8.25}$$

where $\mathbf{J}$ is the current density from origins other than the effective magnetization current in Eq. (3.8.17), e.g., the true currents from flows of free charge carriers or the effective polarization current. With the introduction of the magnetic field vector $\mathbf{H}$ by

$$\mathbf{H} = \frac{1}{\mu_0}\mathbf{B} - \mathbf{M}, \qquad (3.8.26)$$

or

$$\mathbf{B} = \mu_0(\mathbf{H} + \mathbf{M}), \qquad (3.8.27)$$

Eq. (3.8.25) can be written as

$$\nabla \times \mathbf{H} = \mathbf{J}. \qquad (3.8.28)$$

For a linear material,

$$M_k = \chi_{kl}^M H_l. \qquad (3.8.29)$$

Then

$$B_k = \mu_0(H_k + M_k) = \mu_0(\delta_{kl} H_l + \chi_{kl}^M H_l) = \mu_{kl} H_l, \qquad (3.8.30)$$

or

$$\begin{bmatrix} B_1 \\ B_2 \\ B_3 \end{bmatrix} = \begin{bmatrix} \mu_{11} & \mu_{12} & \mu_{13} \\ \mu_{21} & \mu_{22} & \mu_{23} \\ \mu_{31} & \mu_{32} & \mu_{33} \end{bmatrix} \begin{bmatrix} H_1 \\ H_2 \\ H_3 \end{bmatrix}, \qquad (3.8.31)$$

where

$$\mu_{kl} = \mu_0(\delta_{kl} + \chi_{kl}^M), \qquad (3.8.32)$$

is the magnetic permeability of the material. For a linear magnetic material, the energy density is (see Eqs. (3.10.9) and (3.10.10))

$$\hat{U}(\mathbf{B}) = \frac{1}{2}H_i B_i. \qquad (3.8.33)$$

With the following Legendre transform, we introduce a magnetic enthalpy function H by

$$H(\mathbf{H}) = \hat{U} - H_i B_i = -\frac{1}{2}H_i B_i = -\frac{1}{2}\mu_{ij}H_i H_j. \qquad (3.8.34)$$

Then

$$B_i = -\frac{\partial H}{\partial H_i} = \mu_{ik}H_k. \qquad (3.8.35)$$

In a region where $\mathbf{J} = 0$, e.g., a dielectric material in a static state, Eq. (3.8.28) reduces to

$$\nabla \times \mathbf{H} = 0. \qquad (3.8.36)$$

Then a scalar potential ψ can be introduced such that

$$\mathbf{H} = -\nabla \psi, \quad H_i = -\psi_{,i}. \qquad (3.8.37)$$

With ψ, Eq. (3.8.35) becomes

$$B_k = \mu_{kl} H_l = -\mu_{kl} \psi_{,l}, \qquad (3.8.38)$$

which, when substituted into Eq. (3.8.8), yields an equation for ψ:

$$\nabla \cdot \mathbf{B} = B_{k,k} = (\mu_{kl} H_l)_{,k} = -(\mu_{kl} \psi_{,l})_{,k} = 0. \qquad (3.8.39)$$

3.9 Magnetoelectric and Thermal Couplings

Static electric and magnetic fields can interact through magnetoelectric couplings which have various microscopic mechanisms. They are considered macroscopically in what follows, along with thermal couplings. Consider the following free energy F obtained from the energy density $\hat{U}$ (see Eqs. (3.10.9) and (3.10.10)) by a Legendre transform:

$$F(\mathbf{E}, \mathbf{H}, T) = \hat{U}(\mathbf{D}, \mathbf{B}, \eta) - E_i D_i - H_i B_i - T\eta. \qquad (3.9.1)$$

Let Θ_0 be a uniform reference temperature. $\theta(x, t) = T - \Theta_0$ is a small temperature deviation from Θ_0. For linear materials, we expand F at Θ_0 as

$$\begin{aligned}
F = &-\frac{1}{2}\varepsilon_{ij} E_i E_j - \frac{1}{2}\mu_{ij} H_i H_j - \frac{1}{2}\frac{C_V}{\Theta_0}\theta^2 \\
&- m_{ij} E_i H_j - E_i p_i \theta - H_i q_i \theta.
\end{aligned} \qquad (3.9.2)$$

The corresponding constitutive relations are

$$D_i = -\frac{\partial F}{\partial E_i} = \varepsilon_{ik} E_k + m_{ij} H_j + p_i \theta,$$

$$B_i = -\frac{\partial F}{\partial H_i} = \mu_{ik} H_k + m_{ji} E_j + q_i \theta, \qquad (3.9.3)$$

$$\eta = -\frac{\partial F}{\partial \theta} = \frac{C_V}{\Theta_0}\theta + p_i E_i + q_i H_i,$$

where m_{ij} are the magnetoelectric constants, p_i, the pyroelectric constants and q_i, the pyromagnetic constants. Assuming θ is known so that the heat equation is not needed, we have the following equations for the electric and magnetic fields:

$$\mathbf{E} = -\nabla\varphi, \qquad \mathbf{H} = -\nabla\psi, \tag{3.9.4}$$

$$\nabla \cdot \mathbf{D} = \rho^e, \qquad \nabla \cdot \mathbf{B} = 0. \tag{3.9.5}$$

Using Eqs. $(3.9.3)_{1,2}$ and $(3.9.4)$, we can write Eq. $(3.9.5)$ as two equations for φ and ψ.

3.10 Maxwell's Equations

For time-dependent fields, electric and magnetic fields are coupled dynamically and are governed by Maxwell's equations as follows:

$$\nabla \cdot \mathbf{D} = \rho^e,$$

$$\nabla \cdot \mathbf{B} = 0,$$

$$\nabla \times \mathbf{E} = -\frac{\partial \mathbf{B}}{\partial t}, \tag{3.10.1}$$

$$\nabla \times \mathbf{H} = \mathbf{J} + \frac{\partial \mathbf{D}}{\partial t}.$$

Taking the time derivative of the first and the divergence of the fourth of Eq. $(3.10.1)$, respectively, we obtain

$$\frac{\partial}{\partial t}(\nabla \cdot \mathbf{D}) = \frac{\partial \rho^e}{\partial t},$$

$$0 = \nabla \cdot \mathbf{J} + \frac{\partial}{\partial t}(\nabla \cdot \mathbf{D}). \tag{3.10.2}$$

Eliminating the time derivative of the divergence of $\mathbf{D}$, we obtain the conservation of charge as

$$\frac{\partial \rho^e}{\partial t} = -\nabla \cdot \mathbf{J}. \tag{3.10.3}$$

Taking the scalar products of the third and fourth equations of Eq. (3.10.1) with $\mathbf{H}$ and $\mathbf{E}$, respectively, we have

$$\mathbf{H} \cdot (\nabla \times \mathbf{E}) = -\mathbf{H} \cdot \frac{\partial \mathbf{B}}{\partial t},$$

$$\mathbf{E} \cdot (\nabla \times \mathbf{H}) = \mathbf{E} \cdot \mathbf{J} + \mathbf{E} \cdot \frac{\partial \mathbf{D}}{\partial t}. \tag{3.10.4}$$

Subtracting the two equations in Eq. (3.10.4) from each other and using the vector identity

$$\nabla \cdot (\mathbf{E} \times \mathbf{H}) = \mathbf{H} \cdot (\nabla \times \mathbf{E}) - \mathbf{E} \cdot (\nabla \times \mathbf{H}), \tag{3.10.5}$$

we obtain Poynting's theorem

$$\mathbf{E} \cdot \frac{\partial \mathbf{D}}{\partial t} + \mathbf{H} \cdot \frac{\partial \mathbf{B}}{\partial t} + \mathbf{E} \cdot \mathbf{J} = -\nabla \cdot (\mathbf{E} \times \mathbf{H}), \tag{3.10.6}$$

or

$$\frac{\partial \hat{U}}{\partial t} + \mathbf{E} \cdot \mathbf{J} = -\nabla \cdot \mathbf{S}, \tag{3.10.7}$$

where

$$\frac{\partial \hat{U}}{\partial t} = \mathbf{E} \cdot \frac{\partial \mathbf{D}}{\partial t} + \mathbf{H} \cdot \frac{\partial \mathbf{B}}{\partial t}, \quad \mathbf{S} = \mathbf{E} \times \mathbf{H},$$

$$\hat{U} = \hat{U}(\mathbf{D}, \mathbf{B}), \quad \mathbf{E} = \frac{\partial \hat{U}}{\partial \mathbf{D}}, \quad \mathbf{H} = \frac{\partial \hat{U}}{\partial \mathbf{B}}. \tag{3.10.8}$$

$\mathbf{S}$ is the Poynting vector or the electromagnetic energy flux per unit area per unit time. For a linear material,

$$\hat{U}(\mathbf{D}, \mathbf{B}) = \frac{1}{2}\mathbf{E} \cdot \mathbf{D} + \frac{1}{2}\mathbf{H} \cdot \mathbf{B}$$

$$= \frac{1}{2}\mathbf{E} \cdot (\varepsilon_0 \mathbf{E} + \mathbf{P}) + \frac{1}{2}\mathbf{H} \cdot (\mu_0 \mathbf{H} + \mathbf{M}) \tag{3.10.9}$$

$$= \frac{1}{2}\varepsilon_0 \mathbf{E} \cdot \mathbf{E} + \frac{1}{2}\mu_0 \mathbf{H} \cdot \mathbf{H} + \frac{1}{2}\mathbf{E} \cdot \mathbf{P} + \frac{1}{2}\mathbf{H} \cdot \mathbf{M} = U^F + U,$$

where

$$U^F = \frac{1}{2}\varepsilon_0 \mathbf{E} \cdot \mathbf{E} + \frac{1}{2}\mu_0 \mathbf{H} \cdot \mathbf{H},$$

$$U = \frac{1}{2}\mathbf{E} \cdot \mathbf{P} + \frac{1}{2}\mathbf{H} \cdot \mathbf{M}. \tag{3.10.10}$$

Hence, $\hat{U}$ includes the field energy density U^F and the internal energy density U due to polarization and magnetization.

Depending on specific materials, Eq. (3.10.1) may be accompanied by some of the following equations:

$$D_k = \varepsilon_{kl} E_l, \tag{3.10.11}$$

$$B_i = \mu_{ik} H_k, \tag{3.10.12}$$

$$J_k = \sigma_{kl} E_l, \tag{3.10.13}$$

$$J_i^p = qp\mu_{ij}^p E_j - qD_{ij}^p p_{,j}, \tag{3.10.14}$$

$$J_i^n = qn\mu_{ij}^n E_j + qD_{ij}^n n_{,j},$$

$$\rho^e = q(p - n + N_D^+ - N_A^-). \tag{3.10.15}$$

As an example, consider electromagnetic waves in a vacuum. Equation (3.10.1) reduces to

$$\nabla \times \mathbf{E} = -\frac{\partial \mathbf{B}}{\partial t},$$

$$\nabla \times \mathbf{H} = \frac{\partial \mathbf{D}}{\partial t}, \tag{3.10.16}$$

$$\nabla \cdot \mathbf{D} = 0,$$

$$\nabla \cdot \mathbf{B} = 0.$$

For a vacuum,

$$\mathbf{D} = \varepsilon_0 \mathbf{E}, \quad \mathbf{B} = \mu_0 \mathbf{H}. \tag{3.10.17}$$

Taking the curl of Eq. (3.10.16)$_1$, we have

$$\nabla \times (\nabla \times \mathbf{E}) = -\nabla \times \frac{\partial \mathbf{B}}{\partial t}. \tag{3.10.18}$$

With the use of the following vector identity:

$$\nabla \times (\nabla \times \mathbf{E}) = \nabla(\nabla \cdot \mathbf{E}) - \nabla^2 \mathbf{E}, \tag{3.10.19}$$

Eq. (3.10.18) becomes

$$\nabla(\nabla \cdot \mathbf{E}) - \nabla^2 \mathbf{E} = -\frac{\partial}{\partial t}(\nabla \times \mathbf{B}), \tag{3.10.20}$$

or

$$\frac{1}{\varepsilon_0}\nabla(\nabla \cdot \mathbf{D}) - \nabla^2\mathbf{E} = -\frac{\partial}{\partial t}(\nabla \times \mu_0\mathbf{H}), \qquad (3.10.21)$$

where Eq. (3.10.17) has been used. Using Eqs. (3.10.16)$_{2,3}$, we obtain, from Eq. (3.10.21),

$$\mathbf{0} - \nabla^2\mathbf{E} = -\mu_0\frac{\partial}{\partial t}\frac{\partial \mathbf{D}}{\partial t}, \qquad (3.10.22)$$

or

$$\nabla^2\mathbf{E} = \varepsilon_0\mu_0\frac{\partial^2\mathbf{E}}{\partial t^2}. \qquad (3.10.23)$$

Equation (3.10.23) can be written as

$$\nabla^2\mathbf{E} = \frac{1}{c^2}\frac{\partial^2\mathbf{E}}{\partial t^2}, \qquad (3.10.24)$$

which is the standard wave equation where

$$c = \frac{1}{\sqrt{\varepsilon_0\mu_0}}, \qquad (3.10.25)$$

is the wave speed, which in this case is the speed of light in a vacuum. Similarly, it can be shown that

$$\nabla^2\mathbf{B} = \frac{1}{c^2}\frac{\partial^2\mathbf{B}}{\partial t^2}. \qquad (3.10.26)$$

Chapter 4

Piezoelectric and Piezomagnetic Effects

This chapter is on interactions between elastic deformation and electric or magnetic fields. Thermal couplings are also discussed. Three-dimensional theories are presented directly without using one-dimensional models as a transition.

4.1 Piezoelectric Dielectrics

For piezoelectric dielectrics, the equations of motion and the charge equation of electrostatics are from Eqs. (1.6.14) and (3.3.7) [22–25]

$$T_{ji,j} + f_i = \rho \ddot{u}_i,$$
$$D_{i,i} = \rho^e. \tag{4.1.1}$$

The strain–displacement relation and electric field–potential relation are, respectively,

$$S_{ij} = (u_{i,j} + u_{j,i})/2,$$
$$E_i = -\varphi_{,i}. \tag{4.1.2}$$

The constitutive relations are determined by an electric enthalpy density function H defined by

$$H(\mathbf{S}, \mathbf{E}) = \hat{U} - E_i D_i$$
$$= \frac{1}{2} c_{ijkl}^E S_{ij} S_{kl} - e_{ijk} E_i S_{jk} - \frac{1}{2} \varepsilon_{ij}^S E_i E_j, \tag{4.1.3}$$

which produces

$$T_{ij} = \frac{\partial H}{\partial S_{ij}} = c^E_{ijkl} S_{kl} - e_{kij} E_k,$$

$$D_i = -\frac{\partial H}{\partial E_i} = e_{ikl} S_{kl} + \varepsilon^S_{ik} E_k. \tag{4.1.4}$$

The superscript E in the elastic stiffness c^E_{ijkl} indicates that the independent electric constitutive variable is the electric field $\mathbf{E}$. The superscript S in the dielectric constants ε^S_{ij} indicates that the mechanical constitutive variable is the strain tensor $\mathbf{S}$. The material constants in Eq. (4.1.4) have the following symmetries:

$$c^E_{ijkl} = c^E_{jikl} = c^E_{klij},$$

$$e_{kij} = e_{kji}, \quad \varepsilon^S_{ij} = \varepsilon^S_{ji}. \tag{4.1.5}$$

We also assume that the elastic and dielectric material tensors are positive-definite in the following sense:

$$c^E_{ijkl} S_{ij} S_{kl} \geq 0 \quad \text{for any } S_{ij} = S_{ji},$$

$$\text{and } c^E_{ijkl} S_{ij} S_{kl} = 0 \quad \Rightarrow \quad S_{ij} = 0;$$

$$\varepsilon^S_{ij} E_i E_j \geq 0 \quad \text{for any } E_i,$$

$$\text{and } \varepsilon^S_{ij} E_i E_j = 0 \quad \Rightarrow \quad E_i = 0. \tag{4.1.6}$$

With successive substitutions from Eqs. (4.1.4) and (4.1.2), we can write Eq. (4.1.1) as two equations for $\mathbf{u}$ and φ, as follows:

$$c^E_{ijkl} u_{k,lj} + e_{kij} \varphi_{,kj} + f_i = \rho \ddot{u}_i,$$

$$e_{ikl} u_{k,li} - \varepsilon^S_{ij} \varphi_{,ij} = \rho^e. \tag{4.1.7}$$

In the compact matrix notation, Eq. (4.1.4) becomes

$$T_p = c^E_{pq} S_q - e_{kp} E_k,$$

$$D_i = e_{iq} S_q + \varepsilon^S_{ik} E_k, \tag{4.1.8}$$

or

$$
\begin{bmatrix} T_1 \\ T_2 \\ T_3 \\ T_4 \\ T_5 \\ T_6 \end{bmatrix} = \begin{bmatrix} c_{11}^E & c_{12}^E & c_{13}^E & c_{14}^E & c_{15}^E & c_{16}^E \\ c_{21}^E & c_{22}^E & c_{23}^E & c_{24}^E & c_{25}^E & c_{26}^E \\ c_{31}^E & c_{32}^E & c_{33}^E & c_{34}^E & c_{35}^E & c_{36}^E \\ c_{41}^E & c_{42}^E & c_{43}^E & c_{44}^E & c_{45}^E & c_{46}^E \\ c_{51}^E & c_{52}^E & c_{53}^E & c_{54}^E & c_{55}^E & c_{56}^E \\ c_{61}^E & c_{62}^E & c_{63}^E & c_{64}^E & c_{65}^E & c_{66}^E \end{bmatrix} \begin{bmatrix} S_1 \\ S_2 \\ S_3 \\ S_4 \\ S_5 \\ S_6 \end{bmatrix} - \begin{bmatrix} e_{11} & e_{21} & e_{31} \\ e_{12} & e_{22} & e_{32} \\ e_{13} & e_{23} & e_{33} \\ e_{14} & e_{24} & e_{34} \\ e_{15} & e_{25} & e_{35} \\ e_{16} & e_{26} & e_{36} \end{bmatrix} \begin{bmatrix} E_1 \\ E_2 \\ E_3 \end{bmatrix},
$$

$$(4.1.9)$$

and

$$
\begin{bmatrix} D_1 \\ D_2 \\ D_3 \end{bmatrix} = \begin{bmatrix} e_{11} & e_{12} & e_{13} & e_{14} & e_{15} & e_{16} \\ e_{21} & e_{22} & e_{23} & e_{24} & e_{25} & e_{26} \\ e_{31} & e_{32} & e_{33} & e_{34} & e_{35} & e_{36} \end{bmatrix} \begin{bmatrix} S_1 \\ S_2 \\ S_3 \\ S_4 \\ S_5 \\ S_6 \end{bmatrix} + \begin{bmatrix} \varepsilon_{11}^S & \varepsilon_{12}^S & \varepsilon_{13}^S \\ \varepsilon_{21}^S & \varepsilon_{22}^S & \varepsilon_{23}^S \\ \varepsilon_{31}^S & \varepsilon_{32}^S & \varepsilon_{33}^S \end{bmatrix} \begin{bmatrix} E_1 \\ E_2 \\ E_3 \end{bmatrix}.
$$

$$(4.1.10)$$

The constitutive relations in Eq. (4.1.4) can be written in other equivalent forms as follows:

$$
\begin{aligned} S_{ij} &= s_{ijkl}^E T_{kl} + d_{kij} E_k, \\ D_i &= d_{ikl} T_{kl} + \varepsilon_{ik}^T E_k, \end{aligned}
$$

$$(4.1.11)$$

and

$$
\begin{aligned} S_{ij} &= s_{ijkl}^D T_{kl} + g_{kij} D_k, \\ E_i &= -g_{ikl} T_{kl} + \beta_{ik}^T D_k. \end{aligned}
$$

$$(4.1.12)$$

Equations (4.1.11) and (4.1.12) can also be written in matrix forms. The matrices of the material constants in various expressions are related by

$$c_{pr}^E s_{qr}^E = \delta_{pq}, \quad c_{pr}^D s_{qr}^D = \delta_{pq},$$
$$\beta_{ik}^S \varepsilon_{jk}^S = \delta_{ij}, \quad \beta_{ik}^T \varepsilon_{jk}^T = \delta_{ij},$$

$$(4.1.13)$$

$$c_{pq}^D = c_{pq}^E + e_{kp} h_{kq}, \quad s_{pq}^D = s_{pq}^E - d_{kp} g_{kq},$$
$$\varepsilon_{ij}^T = \varepsilon_{ij}^S + d_{iq} e_{jq}, \quad \beta_{ij}^T = \beta_{ij}^S - g_{iq} h_{jq},$$

$$(4.1.14)$$

$$e_{ip} = d_{iq} c_{pq}^E, \quad d_{ip} = \varepsilon_{ik}^T g_{kp},$$
$$g_{ip} = \beta_{ik}^T d_{kp}, \quad h_{ip} = g_{iq} c_{qp}^D.$$

$$(4.1.15)$$

As an example, we begin with Eq. (4.1.8) in matric form as

$$\{T\} = [c^E]\{S\} - [e]^T \{E\},$$
$$\{D\} = [e]\{S\} + [\varepsilon^S]\{E\}.$$

$$(4.1.16)$$

From Eq. $(4.16.1)_1$,

$$[c^E]\{S\} = \{T\} + [e]^T \{E\}.$$

$$(4.1.17)$$

The multiplication of both sides of Eq. (4.1.17) by the inverse matrix of $[c^E]$ yields

$$\{S\} = [c^E]^{-1}\{T\} + [c^E]^{-1}[e]^T \{E\}.$$

$$(4.1.18)$$

The substituting of Eq. (4.1.18) into Eq. $(4.1.16)_2$ gives

$$\{D\} = [e]([c^E]^{-1}\{T\} + [c^E]^{-1}[e]^T \{E\}) + [\varepsilon^S]\{E\}$$
$$= [e][c^E]^{-1}\{T\} + ([e][c^E]^{-1}[e]^T + [\varepsilon^S])\{E\}.$$

$$(4.1.19)$$

Equations (4.1.18) and (4.1.19) can be written as

$$\{S\} = [s^E]\{T\} + [d]^T \{E\},$$
$$\{D\} = [d]\{T\} + [\varepsilon^T]\{E\},$$

$$(4.1.20)$$

or

$$S_p = s_{pq}^E T_q + d_{kp} E_k,$$
$$D_i = d_{iq} T_q + \varepsilon_{ik}^T E_k, \tag{4.1.21}$$

where

$$[s^E] = [c^E]^{-1}, \quad [d] = [e][c^E]^{-1},$$
$$[\varepsilon^T] = [\varepsilon^S] + [e][c^E]^{-1}[e]^T, \tag{4.1.22}$$

or

$$c_{pr}^E s_{qr}^E = \delta_{pq}, \quad e_{ip} = d_{iq} c_{pq}^E,$$
$$\varepsilon_{ij}^T = \varepsilon_{ij}^S + d_{iq} e_{jq}. \tag{4.1.23}$$

As a specific class of piezoelectric materials, for ceramics poled along x_3 or hexagonal crystals of class (6mm) or C_{6v} with the c-axis along x_3, the material matrices are

$$\begin{bmatrix} c_{11} & c_{12} & c_{13} & 0 & 0 & 0 \\ c_{21} & c_{11} & c_{13} & 0 & 0 & 0 \\ c_{31} & c_{31} & c_{33} & 0 & 0 & 0 \\ 0 & 0 & 0 & c_{55} & 0 & 0 \\ 0 & 0 & 0 & 0 & c_{55} & 0 \\ 0 & 0 & 0 & 0 & 0 & c_{66} \end{bmatrix}, \tag{4.1.24}$$

$$\begin{bmatrix} 0 & 0 & 0 & 0 & e_{15} & 0 \\ 0 & 0 & 0 & e_{15} & 0 & 0 \\ e_{31} & e_{31} & e_{33} & 0 & 0 & 0 \end{bmatrix}, \begin{bmatrix} \varepsilon_{11} & 0 & 0 \\ 0 & \varepsilon_{11} & 0 \\ 0 & 0 & \varepsilon_{33} \end{bmatrix}, \tag{4.1.25}$$

where $c_{66} = (c_{11} - c_{12})/2$ and we have neglected superscripts E and S. The corresponding constitutive relations take the following form:

$$T_{11} = c_{11} u_{1,1} + c_{12} u_{2,2} + c_{13} u_{3,3} + e_{31} \varphi_{,3},$$
$$T_{22} = c_{12} u_{1,1} + c_{11} u_{2,2} + c_{13} u_{3,3} + e_{31} \varphi_{,3},$$
$$T_{33} = c_{13} u_{1,1} + c_{13} u_{2,2} + c_{33} u_{3,3} + e_{33} \varphi_{,3}, \tag{4.1.26}$$
$$T_{23} = c_{55}(u_{2,3} + u_{3,2}) + e_{15} \varphi_{,2},$$
$$T_{31} = c_{55}(u_{3,1} + u_{1,3}) + e_{15} \varphi_{,1},$$
$$T_{12} = c_{66}(u_{1,2} + u_{2,1}),$$

and

$$D_1 = e_{15}(u_{3,1} + u_{1,3}) - \varepsilon_{11}\varphi_{,1},$$
$$D_2 = e_{15}(u_{2,3} + u_{3,2}) - \varepsilon_{11}\varphi_{,2},$$
$$D_3 = e_{31}(u_{1,1} + u_{2,2}) + e_{33}u_{3,3} - \varepsilon_{33}\varphi_{,3}.$$

(4.1.27)

The equations of motion and the charge equation of electrostatics are

$$c_{11}u_{1,11} + (c_{12} + c_{66})u_{2,12} + (c_{13} + c_{55})u_{3,13} + c_{66}u_{1,22}$$
$$+ c_{55}u_{1,33} + (e_{31} + e_{15})\varphi_{,13} + f_1 = \rho\ddot{u}_1,$$
$$c_{66}u_{2,11} + (c_{12} + c_{66})u_{1,12} + c_{11}u_{2,22} + (c_{13} + c_{55})u_{3,23}$$
$$+ c_{55}u_{2,33} + (e_{31} + e_{15})\varphi_{,23} + f_2 = \rho\ddot{u}_2,$$
$$c_{55}u_{3,11} + (c_{55} + c_{13})u_{1,31} + c_{55}u_{3,22} + (c_{13} + c_{55})u_{2,23}$$
$$+ c_{33}u_{3,33} + e_{15}(\varphi_{,11} + \varphi_{,22}) + e_{33}\varphi_{,33} + f_3 = \rho\ddot{u}_3,$$
$$e_{15}u_{3,11} + (e_{15} + e_{31})u_{1,13} + e_{15}u_{3,22} + (e_{15} + e_{31})u_{2,32}$$
$$+ e_{33}u_{3,33} - \varepsilon_{11}(\varphi_{,11} + \varphi_{,22}) - \varepsilon_{33}\varphi_{,33} = \rho^e.$$

(4.1.28)

(4.1.29)

Consider longitudinal plane waves described by

$$u_1 = u_2 = 0, \quad u_3 = u_3(x_3,t), \quad \varphi = \varphi(x_3,t).$$ (4.1.30)

Equations (4.1.28) and (4.1.29) reduce to

$$c_{33}u_{3,33} + e_{33}\varphi_{,33} = \rho\ddot{u}_3,$$ (4.1.31)
$$e_{33}u_{3,33} - \varepsilon_{33}\varphi_{,33} = 0.$$

Let

$$\begin{bmatrix} u_3 \\ \varphi \end{bmatrix} = \begin{bmatrix} U \\ \Phi \end{bmatrix} \exp[i(kx_3 - \omega t)].$$ (4.1.32)

The substitution of Eq. (4.1.32) into Eq. (4.1.31) results in

$$c_{33}k^2 U + e_{33}k^2\Phi = \rho\omega^2 U,$$
$$e_{33}k^2 U - \varepsilon_{33}k^2\Phi = 0.$$

(4.1.33)

From Eq. (4.1.33)$_2$,

$$\Phi = \frac{e_{33}}{\varepsilon_{33}} U. \qquad (4.1.34)$$

Substituting Eq. (4.1.34) into Eq. (4.1.33)$_1$, we obtain the wave speed as

$$\frac{\omega}{k} = \sqrt{\frac{c_{33}'}{\rho}}, \qquad (4.1.35)$$

where c_{33}' is a piezoelectrically stiffened elastic constant:

$$c_{33}' = c_{33} \left(1 + \frac{e_{33}^2}{\varepsilon_{33} c_{33}} \right) = c_{33} \left(1 + k_{33}^2 \right),$$

$$k_{33}^2 = \frac{e_{33}^2}{\varepsilon_{33} c_{33}}. \qquad (4.1.36)$$

Similarly, for transverse plane waves described by

$$u_1 = u_2 = 0, \quad u_3 = u_3(x_1, t), \quad \varphi = \varphi(x_1, t), \qquad (4.1.37)$$

Eqs. (4.1.28) and (4.1.29) reduce to

$$c_{55} u_{3,11} + e_{15} \varphi_{,11} = \rho \ddot{u}_3,$$

$$e_{15} u_{3,11} - \varepsilon_{11} \varphi_{,11} = 0. \qquad (4.1.38)$$

Let

$$\begin{bmatrix} u_3 \\ \varphi \end{bmatrix} = \begin{bmatrix} U \\ \Phi \end{bmatrix} \exp[i(kx_1 - \omega t)]. \qquad (4.1.39)$$

Equation (4.1.38) becomes

$$c_{55} k^2 U + e_{15} k^2 \Phi = \rho \omega^2 U,$$

$$e_{15} k^2 U - \varepsilon_{11} k^2 \Phi = 0. \qquad (4.1.40)$$

Eliminating Φ and U, we obtain the transverse or shear wave speed as

$$\frac{\omega}{k} = \sqrt{\frac{c'_{55}}{\rho}}, \qquad (4.1.41)$$

where

$$c'_{55} = c_{55}\left(1 + \frac{e_{15}^2}{\varepsilon_{11}c_{55}}\right) = c_{55}\left(1 + k_{15}^2\right), \qquad (4.1.42)$$

$$k_{15}^2 = \frac{e_{15}^2}{\varepsilon_{11}c_{55}}.$$

Cubic crystals of class (43m) are also piezoelectric. Their material matrices are

$$[c_{pq}] = \begin{bmatrix} c_{11} & c_{12} & c_{12} & 0 & 0 & 0 \\ c_{12} & c_{11} & c_{12} & 0 & 0 & 0 \\ c_{12} & c_{12} & c_{11} & 0 & 0 & 0 \\ 0 & 0 & 0 & c_{44} & 0 & 0 \\ 0 & 0 & 0 & 0 & c_{44} & 0 \\ 0 & 0 & 0 & 0 & 0 & c_{44} \end{bmatrix}, \qquad (4.1.43)$$

$$[e_{ip}] = \begin{bmatrix} 0 & 0 & 0 & e_{14} & 0 & 0 \\ 0 & 0 & 0 & 0 & e_{14} & 0 \\ 0 & 0 & 0 & 0 & 0 & e_{14} \end{bmatrix}, \quad [\varepsilon_{ij}] = \begin{bmatrix} \varepsilon_{11} & 0 & 0 \\ 0 & \varepsilon_{11} & 0 \\ 0 & 0 & \varepsilon_{11} \end{bmatrix}. $$
$$(4.1.44)$$

4.2   Thermopiezoelectric Dielectrics

The field equations for the theory of thermopiezoelectricity are Eqs. (4.1.1) and (2.4.18)$_2$ [25–28]

$$T_{ji,j} + f_i = \rho\ddot{u}_i,$$
$$D_{i,i} = \rho^e, \qquad (4.2.1)$$
$$\Theta_0\dot{\eta} = r - h_{i,i}.$$

Let Θ_0 be a uniform reference temperature. $\theta(x,t) = T - \Theta_0$ is a small temperature deviation from Θ_0. For linear materials, we construct the free energy density F from Eqs. (2.4.14), (3.9.2) and (4.1.3) and expand F at Θ_0 as

$$F(\mathbf{S}, \mathbf{E}, \theta) = \hat{U}(\mathbf{S}, \mathbf{D}, \eta) - T\eta - E_k D_k$$

$$= \frac{1}{2} c_{ijkl} S_{ij} S_{kl} - e_{kij} E_k S_{ij} - \frac{1}{2} \varepsilon_{ij} E_i E_j \qquad (4.2.2)$$

$$- \frac{1}{2} \frac{C_V}{\Theta_0} \theta^2 - \lambda_{kl} S_{kl} \theta - p_k E_k \theta,$$

which leads to the following constitutive relations:

$$T_{ij} = \frac{\partial F}{\partial S_{ij}} = c_{ijkl} S_{kl} - e_{kij} E_k - \lambda_{ij}\theta,$$

$$D_i = -\frac{\partial F}{\partial E_i} = e_{ijk} S_{jk} + \varepsilon_{ij} E_j + p_k\theta, \qquad (4.2.3)$$

$$\eta = -\frac{\partial F}{\partial \theta} = \lambda_{kl} S_{kl} + p_k E_k + \frac{C_V}{\Theta_0}\theta,$$

where

$$S_{ij} = (u_{i,j} + u_{j,i})/2,$$
$$E_i = -\varphi_{,i}. \qquad (4.2.4)$$

In addition, we have Fourier's law for the heat flux vector $\mathbf{h}$ as follows:

$$h_k = -\kappa_{kl}\theta_{,l}. \qquad (4.2.5)$$

With successive substitutions from Eqs. (4.2.3)–(4.2.5), we can write Eq. (4.2.1) as five equations for $\mathbf{u}$, φ and θ.

Similar to Eqs. (4.1.11) and (4.1.12), constitutive relations in other forms than Eq. (4.2.3) with the stress tensor $\mathbf{T}$ and/or the electric displacement vector $\mathbf{D}$ as constitutive arguments can be derived.

4.3 Piezoelectric Semiconductors

For piezoelectric semiconductors, the field equations are from Eqs. (4.1.1)$_1$, (3.7.1)$_3$ and (3.7.2) [29–32]

$$T_{ji,j} + f_i = \rho \ddot{u}_i,$$

$$D_{i,i} = \rho^e = q(p - n + N_D^+ - N_A^-), \qquad (4.3.1)$$

$$q\dot{p} = -J^p_{i,i} + \gamma^p,$$
$$q\dot{n} = J^n_{i,i} + \gamma^n. \tag{4.3.2}$$

The following constitutive relations are from Eqs. (4.1.4) and (3.7.3):

$$T_{ij} = c_{ijkl}S_{kl} - e_{kij}E_k,$$
$$D_i = e_{ijk}S_{jk} + \varepsilon_{ij}E_j, \tag{4.3.3}$$

$$J^p_i = qp\mu^p_{ij}E_j - qD^p_{ij}p_{,j},$$
$$J^n_i = qn\mu^n_{ij}E_j + qD^n_{ij}n_{,j}. \tag{4.3.4}$$

We also have the strain–displacement and electric field–potential relations:

$$S_{ij} = (u_{i,j} + u_{j,i})/2,$$
$$E_i = -\varphi_{,i}. \tag{4.3.5}$$

With substitutions from Eqs. (4.3.3)–(4.3.5), we can write Eqs. (4.3.1) and (4.3.2) as six equations for $\mathbf{u}$, φ, p and n.

We assume uniform p_0 and n_0 at a reference state with

$$p = N^-_A = p_0, \quad n = N^+_D = n_0,$$
$$\mathbf{T} = 0, \quad \mathbf{S} = 0, \tag{4.3.6}$$
$$\mathbf{D} = 0, \quad \mathbf{E} = 0, \quad \mathbf{J}^p = 0, \quad \mathbf{J}^n = 0.$$

Let

$$p = p_0 + \Delta p,$$
$$n = n_0 + \Delta n. \tag{4.3.7}$$

For small Δp and Δn, Eq. (4.3.4) may be linearized as

$$J^p_i \cong qp_0\mu^p_{ij}E_j - qD^p_{ij}(\Delta p)_{,j},$$
$$J^n_i \cong qn_0\mu^n_{ij}E_j + qD^n_{ij}(\Delta n)_{,j}. \tag{4.3.8}$$

With Δp and Δn, Eqs. (4.3.1) and (4.3.2) become

$$T_{ji,j} + f_i = \rho \ddot{u}_i,$$
$$D_{i,i} = q(\Delta p - \Delta n), \tag{4.3.9}$$

$$q\frac{\partial(\Delta p)}{\partial t} = -J^p_{i,i} + \gamma^p,$$
$$q\frac{\partial(\Delta n)}{\partial t} = J^n_{i,i} + \gamma^n, \tag{4.3.10}$$

which can be written as six linear equations for $\mathbf{u}$, φ, Δp and Δn.

As an example, consider the propagation of plane waves [33] in an n-type hexagonal crystal described by

$$u_3 = u_3(x_3, t), \quad u_1 = u_2 = 0, \quad \varphi = \varphi(x_3, t),$$
$$\Delta n = \Delta n(x_3, t), \quad p_0 = 0, \quad \Delta p = 0. \tag{4.3.11}$$

The analysis in [33] is nonlinear. A linear analysis is given as follows. For the fields in Eq. (4.3.11), we have

$$S_{33} = u_{3,3}, \quad E_3 = -\varphi_{,3}, \tag{4.3.12}$$

$$T_{33} = c_{33}S_{33} - e_{33}E_3 = c_{33}u_{3,3} + e_{33}\varphi_{,3},$$
$$D_3 = e_{33}S_{33} + \varepsilon_{33}E_3 = e_{33}u_{3,3} - \varepsilon_{33}\varphi_{,3}, \tag{4.3.13}$$

$$J^n_3 = qn_0\mu^n_{33}E_3 + qD^n_{33}(\Delta n)_{,3}$$
$$= -qn_0\mu^n_{33}\varphi_{,3} + qD^n_{33}(\Delta n)_{,3}. \tag{4.3.14}$$

The governing equations for u_3, φ and Δn are

$$T_{33,3} = c_{33}u_{3,33} + e_{33}\varphi_{,33} = \rho\ddot{u}_3,$$
$$D_{3,3} = e_{33}u_{3,33} - \varepsilon_{33}\varphi_{,33} = -q(\Delta n), \tag{4.3.15}$$

$$\frac{\partial}{\partial t}(\Delta n) = J^n_{3,3} = -n_0\mu^n_{33}\varphi_{,33} + D^n_{33}(\Delta n)_{,33}. \tag{4.3.16}$$

Let

$$\left\{ \begin{array}{c} u_3 \\ \varphi \\ \Delta n \end{array} \right\} = \left\{ \begin{array}{c} U \\ \Phi \\ N \end{array} \right\} \exp[i(kx_3 - \omega t)]. \tag{4.3.17}$$

Substituting Eq. (4.3.17) into Eqs. (4.3.15) and (4.3.16), we have

$$c_{33}k^2 U + e_{33}k^2 \Phi = \rho\omega^2 U,$$
$$e_{33}k^2 U - \varepsilon_{33}k^2 \Phi = qN, \qquad (4.3.18)$$
$$-i\omega N = n_0 \mu_{33}^n k^2 \Phi - D_{33}^n k^2 N.$$

From Eq. $(4.3.18)_3$, we obtain

$$N = \frac{n_0 \mu_{33}^n k^2}{D_{33}^n k^2 - i\omega} \Phi. \qquad (4.3.19)$$

Then, from Eqs. $(4.3.18)_2$ and $(4.3.19)$,

$$\Phi = \frac{e_{33}k^2 (D_{33}^n k^2 - i\omega)}{q n_0 \mu_{33}^n k^2 + \varepsilon_{33}k^2 (D_{33}^n k^2 - i\omega)} U. \qquad (4.3.20)$$

With Eq. (4.3.20), we obtain the wave speed from Eq. $(4.3.18)_1$ as

$$\frac{\omega}{k} = \sqrt{\frac{c'_{33}}{\rho}}, \qquad (4.3.21)$$

where c'_{33} is an effective elastic constant:

$$c'_{33} = c_{33} \left[1 + \frac{e_{33}^2}{c_{33}\varepsilon_{33}} \frac{\varepsilon_{33}k^2 (D_{33}^n k^2 - i\omega)}{q n_0 \mu_{33}^n k^2 + \varepsilon_{33}k^2 (D_{33}^n k^2 - i\omega)} \right]. \qquad (4.3.22)$$

Equation (4.3.22) shows the familiar piezoelectric stiffening effect through $e^2/(c\varepsilon)$. c'_{33} is complex, suggesting dissipation due to semi-conduction. Equation (4.3.22) also shows dispersion because of its dependence on ω.

4.4　Piezomagnetic Dielectrics

For piezomagnetic dielectrics that are also piezoelectric, the following field equations are from Eqs. (4.1.1) and (3.8.8):

$$T_{ji,j} + f_i = \rho \ddot{u}_i,$$
$$D_{i,i} = \rho^e, \qquad (4.4.1)$$
$$B_{i,i} = 0.$$

The enthalpy function H is a generalization of Eqs. (4.1.3) to include some of the magnetic effects in Eq. (3.9.2):

$$H(\mathbf{S}, \mathbf{E}, \mathbf{H}) = \hat{U}(\mathbf{S}, \mathbf{D}, \mathbf{B}) - E_k D_k - H_k B_k$$

$$= \frac{1}{2} c_{ijkl} S_{ij} S_{kl} - e_{ijk} E_i S_{jk} - h_{ijk} H_i S_{jk} \qquad (4.4.2)$$

$$- \frac{1}{2} \varepsilon_{ij} E_i E_j - \frac{1}{2} \mu_{ij} H_i H_j - m_{ij} E_i H_j,$$

where h_{ijk} are piezomagnetic constants and m_{ij}, magnetoelectric constants. The constitutive relations determined by H are

$$T_{ij} = \frac{\partial H}{\partial S_{ij}} = c_{ijkl} S_{kl} - e_{kij} E_k - h_{kij} H_k,$$

$$D_i = -\frac{\partial H}{\partial E_i} = e_{ikl} S_{kl} + \varepsilon_{ik} E_k + m_{ij} H_j, \qquad (4.4.3)$$

$$B_i = -\frac{\partial H}{\partial H_i} = h_{ikl} S_{kl} + m_{ji} E_j + \mu_{ik} H_k.$$

In addition,

$$S_{ij} = (u_{i,j} + u_{j,i})/2,$$
$$E_i = -\varphi_{,i}, \quad H_i = -\psi_{,i}. \qquad (4.4.4)$$

With substitutions from Eqs. (4.4.3) and (4.4.4), we can write Eq. (4.4.1) as five equations for $\mathbf{u}$, φ and ψ. The piezomagnetic materials to be considered in the rest of the book are assumed to have the following material matrices:

$$\begin{bmatrix} c_{11} & c_{12} & c_{13} & 0 & 0 & 0 \\ c_{21} & c_{11} & c_{13} & 0 & 0 & 0 \\ c_{31} & c_{31} & c_{33} & 0 & 0 & 0 \\ 0 & 0 & 0 & c_{55} & 0 & 0 \\ 0 & 0 & 0 & 0 & c_{55} & 0 \\ 0 & 0 & 0 & 0 & 0 & c_{66} \end{bmatrix}, \qquad (4.4.5)$$

$$\begin{bmatrix} 0 & 0 & 0 & 0 & h_{15} & 0 \\ 0 & 0 & 0 & h_{15} & 0 & 0 \\ h_{31} & h_{31} & h_{33} & 0 & 0 & 0 \end{bmatrix}, \begin{bmatrix} \mu_{11} & 0 & 0 \\ 0 & \mu_{11} & 0 \\ 0 & 0 & \mu_{33} \end{bmatrix}, \quad (4.4.6)$$

where $c_{66} = (c_{11} - c_{12})/2$.

4.5 Gibbs Free Energy

The thermodynamic functions we have introduced so far such as the energy density $\hat{U}$, the Helmholtz free energy F and the enthalpy function H all have the strain tensor $\mathbf{S}$ as the independent mechanical constitutive variable. In the development of the one-dimensional theories of thin rods in Chapters 5–7, since many stress components are approximately equal to zero, it is more convenient to use the stress tensor $\mathbf{T}$ as the independent mechanical constitutive variable. Therefore, we introduce the following Gibbs free energy G:

$$G(\mathbf{T}, \mathbf{E}, \mathbf{H}, T) = \hat{U}(\mathbf{S}, \mathbf{D}, \mathbf{B}, \eta) - T_{kl}S_{kl} - E_k D_k - H_k B_k - T\eta. \quad (4.5.1)$$

We expand G at a uniform reference temperature Θ_0 and denote $\theta = T - \Theta_0$. Keeping second-order terms only, we have

$$G(\mathbf{T}, \mathbf{E}, \mathbf{H}, \theta) = -\frac{1}{2}s_{ijkl}^{EH\theta}T_{ij}T_{kl} - d_{ijk}^{H\theta}E_iT_{jk} - h_{ijk}^{E\theta}H_iT_{jk}$$

$$- \frac{1}{2}\varepsilon_{ij}^{TH\theta}E_iE_j - \frac{1}{2}\mu_{ij}^{TE\theta}H_iH_j - m_{ij}^{T\theta}E_iH_j \quad (4.5.2)$$

$$- \frac{1}{2}\frac{C^{TEH}}{\Theta_0}\theta^2 - \alpha_{kl}^{EH}T_{kl}\theta - p_k^{TH}E_k\theta - q_k^{TE}H_k\theta.$$

The superscripts T, E, H and θ in the material constants indicate stress, electric field, magnetic field and temperature, respectively. The constitutive relations generated by Eq. (4.5.2) are

$$S_{ij} = -\frac{\partial G}{\partial T_{ij}} = s_{ijkl}^{EH\theta}T_{kl} + d_{kij}^{H\theta}E_k + h_{kij}^{E\theta}H_k + \alpha_{ij}^{EH}\theta, \quad (4.5.3)$$

$$D_i = -\frac{\partial G}{\partial E_i} = d_{ijk}^{H\theta} T_{jk} + \varepsilon_{ij}^{TH\theta} E_j + m_{ij}^{T\theta} H_j + p_i^{TH}\theta, \quad (4.5.4)$$

$$B_i = -\frac{\partial G}{\partial H_i} = h_{ijk}^{E\theta} T_{jk} + m_{ji}^{T\theta} E_j + \mu_{ij}^{TE\theta} H_j + q_i^{TE}\theta, \quad (4.5.5)$$

$$\eta = -\frac{\partial G}{\partial \theta} = \alpha_{kl}^{EH} T_{kl} + p_k^{TH} E_k + q_k^{TE} H_k + \frac{C^{TEH}}{\Theta_0}\theta. \quad (4.5.6)$$

In the rest of this book, Eqs. (4.5.3)–(4.5.6) or parts of them are used often. The superscripts in the material constants will be dropped when there is no ambiguity.

Chapter 5

Extension of Rods

In this chapter, we derive one-dimensional equations for extensional motions of thin rods from the three-dimensional equations. The chapter begins with piezoelectric dielectric rods including composite rods with piezoelectric and elastic dielectric layers. Then it progresses to piezoelectric semiconductor rods and piezomagnetic dielectric as well as semiconductor rods.

5.1 Piezoelectric Dielectric Rods with Transverse Poling

Consider the rectangular ceramic rod of length L, width w and thickness h as shown in Fig. 5.1, where $L \gg w \gg h$. The poling direction is along x_3 [23]. The cross-sectional area is $A = wh$.

The following equations are in fact valid for other piezoelectric dielectrics in general, not limited to polarized ceramics. As an approximation, it is appropriate to take the vanishing boundary stresses on the surfaces bounding the two small dimensions to vanish everywhere (stress relaxation for thin rods). Consequently,

$$T_{11} \cong T_1(x_1, t), \qquad \text{and all other } T_{ij} \cong 0. \qquad (5.1.1)$$

When the top and bottom surfaces of the area lw are fully electroded with a driving voltage V across the electrodes ($\varphi = 0$ at the bottom and $\varphi = V$ at the top), the appropriate electrical approximations are

$$E_1 \cong 0, \quad E_2 \cong 0, \quad E_3 \cong -\frac{V}{h}. \qquad (5.1.2)$$

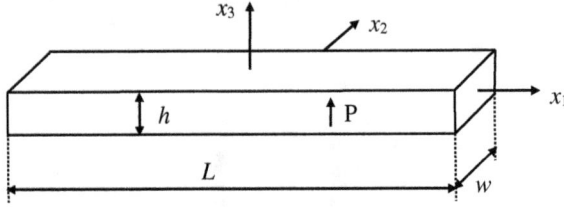

Fig. 5.1.   A ceramic rod with a rectangular cross-section.

The pertinent constitutive relations are

$$S_1 = s_{11}T_1 + d_{31}E_3,$$
$$D_3 = d_{31}T_1 + \varepsilon_{33}E_3, \tag{5.1.3}$$

where $S_1 = u_{1,1}$ and

$$s_{33} = s_{33}^E, \quad \varepsilon_{33} = \varepsilon_{33}^T. \tag{5.1.4}$$

Equation (5.1.3) can be inverted to give

$$T_1 = \bar{c}_{11}S_1 - \bar{e}_{31}E_3 = \bar{c}_{11}u_{1,1} + \bar{e}_{31}\frac{V}{h},$$
$$D_3 = \bar{e}_{31}S_1 + \bar{\varepsilon}_{33}E_3 = \bar{e}_{31}u_{1,1} - \bar{\varepsilon}_{33}\frac{V}{h}, \tag{5.1.5}$$

where the effective one-dimensional material constants are defined by

$$\bar{c}_{11} = 1/s_{11}, \quad \bar{e}_{31} = d_{31}/s_{11},$$
$$\bar{\varepsilon}_{33} = \varepsilon_{33}(1 - k_{31}^2), \quad k_{31}^2 = d_{31}^2/(\varepsilon_{33}s_{11}). \tag{5.1.6}$$

The axial force N over a cross-section is given by

$$N = T_1 A = \bar{c}_{11}Au_{1,1} + \bar{e}_{31}A\frac{V}{h}. \tag{5.1.7}$$

The equation of motion is obtained by applying Newton's second law to the differential element in Fig. 5.2 in the axial direction as follows:

$$N_{,1} + F_1 = \rho A\ddot{u}_1, \tag{5.1.8}$$

which can be written as an equation for u_1. Since the electric field is already known, the charge equation of electrostatics is not needed.

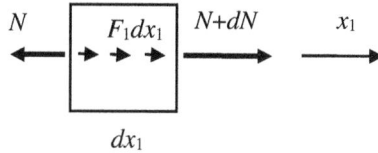

Fig. 5.2. A differential element of the rod under mechanical loads.

As an example, consider a homogeneous rod within $|x_1| < L/2$ under V only. We have the following equation and boundary conditions:

$$N_{,1} = \bar{c}_{11} A u_{1,11} = \rho A \ddot{u}_1, \quad -L/2 < x_1 < L/2,$$

$$N = \bar{c}_{11} A u_{1,1} + \bar{e}_{31} A \frac{V}{h} = 0, \quad x_1 = \pm L/2.$$

(5.1.9)

Equation $(5.1.9)_1$ is the standard wave equation. Equation $(5.1.9)_2$ shows that the applied thickness voltage effectively acts like two extensional end forces on the rod.

In the special case of statics, Eq. (5.1.9) implies that $u_{1,11} = 0$ and $S_1 = u_{1,1} = -\bar{e}_{31} V/(\bar{c}_{11} h)$. The strain and electric field are uniform. If the elongation of the rod is denoted by Δ, the axial strain can be written as $S_1 = \Delta/L$ and the constitutive relations in Eq. (5.1.5) take the following form [34,35]:

$$T_1 = \bar{c}_{11} \frac{\Delta}{L} + \bar{e}_{31} \frac{V}{h},$$

$$D_3 = \bar{e}_{31} \frac{\Delta}{L} - \bar{\varepsilon}_{33} \frac{V}{h},$$

(5.1.10)

which effectively represent a "piezoelectric spring". In the special case when $V = 0$, Eq. $(5.1.10)_1$ reduces to the familiar form of $\Delta = NL/(EA)$ for an elastic rod in a typical mechanics of materials book where $E = \bar{c}_{11}$ is Young's modulus.

For free vibrations of the rod, the electrodes are shorted and $V = 0$. We look for a stationary wave solution in the form of

$$u_1(x_1, t) = u_1(x_1) \exp(i\omega t).$$

(5.1.11)

Then, from Eq. (5.1.9), the eigenvalue problem is

$$\bar{c}_{11} u_{1,11} + \rho \omega^2 u_1, \quad -L/2 < x_1 < L/2,$$

$$u_{1,1} = 0, \quad x_1 = \pm L/2.$$

(5.1.12)

The solution with $\omega = 0$ and $u_1 = $ constant represents a rigid-body mode. For the rest of the modes, we try $u_1 = \sin kx_1$. Then, from Eq. (5.1.11)$_1$,

$$k = \omega\sqrt{\frac{\rho}{\bar{c}_{11}}}. \tag{5.1.13}$$

To satisfy Eq. (5.1.12)$_2$, we must have

$$\cos k\frac{L}{2} = 0, \quad \Rightarrow \quad k^{(n)}\frac{L}{2} = \frac{n\pi}{2}, \quad n = 1,3,5,\ldots, \tag{5.1.14}$$

or

$$\omega^{(n)}\sqrt{\frac{\rho}{\bar{c}_{11}}}\frac{L}{2} = \frac{n\pi}{2}, \quad \omega^{(n)} = \frac{n\pi}{L}\sqrt{\frac{\bar{c}_{11}}{\rho}}, \quad n = 1,3,5,\ldots. \tag{5.1.15}$$

Similarly, by considering $u_1 = \cos kx_1$, the following resonance frequencies can be determined:

$$\omega^{(n)} = \frac{n\pi}{L}\sqrt{\frac{\bar{c}_{11}}{\rho}}, \quad n = 2,4,6,\ldots. \tag{5.1.16}$$

The frequencies in Eqs. (5.1.15) and (5.1.16) are integral multiples of $\omega^{(1)}$ and are called harmonics. $\omega^{(1)}$ is called the fundamental and the rest are called the overtone frequencies. For the one-dimensional model to be valid, the wavelength of the modes should be much larger than the width and thickness of the rod. The one-dimensional model is less accurate for higher-order modes with shorter wavelengths. If a traveling wave $u_1 = \sin(kx_1 - \omega t)$ is substituted into Eq. (5.1.9)$_1$, we still obtain Eq. (5.1.13) which determines the wave speed as $\omega/k = (\bar{c}_{11}/\rho)^{1/2} = (E/\rho)^{1/2}$, which indicates a nondispersive wave.

If the surfaces of the area lh are fully electroded with a driving voltage V across the electrodes, the appropriate electrical approximations are

$$E_1 \cong 0, \quad D_3 \cong 0, \quad E_2 \cong -\frac{V}{w}. \tag{5.1.17}$$

The pertinent constitutive relations are

$$S_1 = s_{11}T_1 + d_{21}E_2 + d_{31}E_3,$$
$$D_2 = d_{21}T_1 + \varepsilon_{22}E_2 + \varepsilon_{23}E_3. \tag{5.1.18}$$

From the boundary conditions on the areas of lw, we take the following to be approximately true everywhere:

$$D_3 = d_{31}T_1 + \varepsilon_{32}E_2 + \varepsilon_{33}E_3 \cong 0. \tag{5.1.19}$$

With Eq. (5.1.19), we can write Eq. (5.1.18) as

$$S_1 = \tilde{s}_{11}T_1 + \tilde{d}_{21}E_2,$$
$$D_2 = \tilde{d}_{21}T_1 + \tilde{\varepsilon}_{22}E_2, \tag{5.1.20}$$

where

$$\tilde{s}_{11} = s_{11} - d_{31}^2/\varepsilon_{33},$$
$$\tilde{d}_{21} = d_{21} - d_{31}\varepsilon_{23}/\varepsilon_{33}, \tag{5.1.21}$$
$$\tilde{\varepsilon}_{22} = \varepsilon_{22} - \varepsilon_{23}^2/\varepsilon_{33}.$$

If the surfaces of areas lw and lh are not electroded, the appropriate electrical approximations are

$$D_2 \cong 0, \quad D_3 \cong 0. \tag{5.1.22}$$

In this case, the pertinent constitutive relations are

$$S_1 = s_{11}T_1 + g_{11}D_1,$$
$$E_1 = -g_{11}T_1 + \beta_{11}D_1. \tag{5.1.23}$$

5.2 Piezoelectric Dielectric Rods with Axial Poling

Consider a cylindrical rod made from polarized ceramics with axial poling (see Fig. 5.3) [23]. The shape of the cross-section can be arbitrary, but its characteristic dimension has to be small compared to the length of the rod. The cylindrical surface is unelectroded. The electric field in the surrounding free space is neglected. The two end faces are electroded. The rod can be excited into extensional deformation or vibration through e_{33} if a voltage is applied across the two end electrodes.

For the extensional motion of the rod, we make the following approximations for uniaxial stress and uniaxial electric displacement:

$$T_{33} \cong T_{33}(x_3, t), \quad \text{all other } T_{ij} \cong 0,$$
$$D_3 \cong D_3(x_3, t), \quad D_1 \cong 0, \quad D_2 \cong 0. \tag{5.2.1}$$

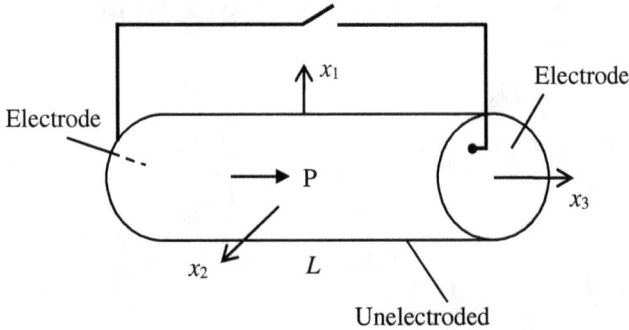

Fig. 5.3.   An axially poled ceramic rod.

Thus, the electromechanical boundary conditions on the lateral surface are satisfied. The relevant constitutive relations take the following form:

$$S_3 = s_{33}T_3 + d_{33}E_3,$$
$$D_3 = d_{33}T_3 + \varepsilon_{33}E_3, \tag{5.2.2}$$

where

$$S_3 = u_{3,3}, \quad E_3 = -\varphi_{,3},$$
$$s_{33} = s_{33}^E, \quad \varepsilon_{33} = \varepsilon_{33}^T. \tag{5.2.3}$$

From Eqs. (5.2.2) and (5.2.3), we obtain

$$T_3 = \bar{c}_{33}S_3 - \bar{e}_{33}E_3 = \bar{c}_{33}u_{3,3} + \bar{e}_{33}\varphi_{,3},$$
$$D_3 = \bar{e}_{33}S_3 + \bar{\varepsilon}_{33}E_3 = \bar{e}_{33}u_{3,3} - \bar{\varepsilon}_{33}\varphi_{,3}, \tag{5.2.4}$$

where the effective one-dimensional material constants are given by

$$\bar{c}_{33} = \frac{1}{s_{33}}, \quad \bar{e}_{33} = \frac{d_{33}}{s_{33}},$$
$$\bar{\varepsilon}_{33} = \varepsilon_{33}\left[1 - \left(k_{33}^l\right)^2\right], \quad \left(k_{33}^l\right)^2 = \frac{d_{33}^2}{\varepsilon_{33}s_{33}}. \tag{5.2.5}$$

The extensional resultant N and the total electric displacement $\hat{D}_3$ over a cross-section with a constant area A are given by

$$N = T_3A = \bar{c}_{33}Au_{3,3} + \bar{e}_{33}A\varphi_{,3},$$
$$\hat{D}_3 = D_3A = \bar{e}_{33}Au_{3,3} - \bar{\varepsilon}_{33}A\varphi_{,3}. \tag{5.2.6}$$

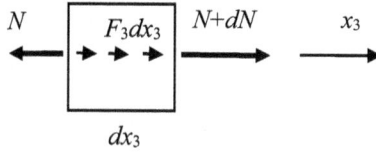

Fig. 5.4.   A differential element of the rod under mechanical loads.

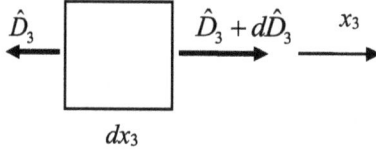

Fig. 5.5.   A differential element of the rod under electrical loads.

The equation of motion is obtained by applying Newton's second law to the differential element of the rod in Fig. 5.4. The charge equation of electrostatics can be obtained by considering the same element under electrical loads as shown in Fig. 5.5. The results are

$$N_{,3} + F_3 = \rho A \ddot{u}_3,$$
$$\hat{D}_{3,3} = 0.$$
$$(5.2.7)$$

Substituting Eq. (5.2.6) into Eq. (5.2.7), for a homogeneous rod, we obtain the following two equations for u_3 and φ:

$$\bar{c}_{33} u_{3,33} + \bar{e}_{33} \varphi_{,33} = \rho \ddot{u}_3,$$
$$\bar{e}_{33} u_{3,33} - \bar{\varepsilon}_{33} \varphi_{,33} = 0.$$
$$(5.2.8)$$

As an example, consider the static extension of a finite rod under equal and opposite end forces $N = pA$ or end tractions $T_3 = p$. Equation (5.2.8) implies that both S_3 and E_3 are constants along the rod. When the end electrodes are shorted, there is no potential difference between the end electrodes. Hence,

$$E_3 \equiv 0,$$ $$(5.2.9)$$

which implies, from Eq. (5.2.2), that

$$D_3 = d_{33}p, \quad S_3 = s_{33}^E p.$$ $$(5.2.10)$$

The mechanical work done to the rod per unit volume during the static extensional process is

$$W_1 = \frac{1}{2}pS_3 = \frac{1}{2}s_{33}^E p^2. \tag{5.2.11}$$

When the end electrodes are open and there are no free charges on the electrodes, $D_3 = 0$ on the end electrodes. Since D_3 is constant along the rod, we have

$$D_3 \equiv 0, \tag{5.2.12}$$

which implies, from Eq. (5.2.2), that

$$E_3 = -\frac{d_{33}}{\varepsilon_{33}^T}p,$$

$$S_3 = s_{33}^E p - d_{33}\frac{d_{33}}{\varepsilon_{33}^T}p = s_{33}^E \left(1 - \frac{d_{33}^2}{\varepsilon_{33}^T s_{33}^E}\right)p. \tag{5.2.13}$$

In this case, the mechanical work done to the rod per unit volume is

$$W_2 = \frac{1}{2}pS_3 = \frac{1}{2}s_{33}^E \left(1 - \frac{d_{33}^2}{\varepsilon_{33}^T s_{33}^E}\right)p^2 = \frac{1}{2}s_{33}^E \left[1 - (k_{33}^l)^2\right]p^2. \tag{5.2.14}$$

We assume that

$$\frac{d_{33}^2}{\varepsilon_{33}^T s_{33}^E} > 0. \tag{5.2.15}$$

Then the extensional strain S_3 given by Eq. (5.2.10) is larger than that given by Eq. (5.2.13). As a consequence, we have

$$W_1 > W_2. \tag{5.2.16}$$

Therefore, the rod appears to be stiffer when the electrodes are open and an axial electric field/polarization is produced. This is called the piezoelectric stiffening effect. Graphically, W_1 and W_2 are represented by the areas in Fig. 5.6. k_{33}^l is called the longitudinal electromechanical coupling factor. It can be written as

$$(k_{33}^l)^2 = \frac{W_1 - W_2}{W_1} = \frac{d_{33}^2}{\varepsilon_{33}^T s_{33}^E}. \tag{5.2.17}$$

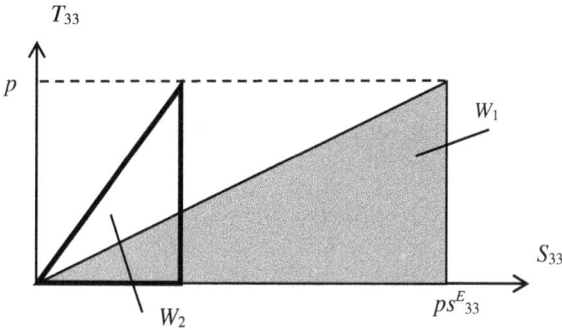

Fig. 5.6. Work done to the ceramic rod per unit volume along different paths.

As a numerical example, for PZT-5H, a common ceramic, from the material constants in Appendix 2,

$$(k_{33}^l)^2 = \frac{(593 \times 10^{-12})^2}{(3400 \times 8.85 \times 10^{-12})(20.7 \times 10^{-12})} = 0.56, \tag{5.2.18}$$

$$k_{33}^l = 0.75,$$

which is typical for polarized ceramics.

As another example, consider the free vibration of a finite rod within $|x_3| < L/2$ with open end electrodes. The electrical end conditions are

$$\hat{D}_3 = 0, \quad x_3 = \pm L/2. \tag{5.2.19}$$

Equation $(5.2.7)_2$ implies that $\hat{D}_3$ is independent of x_3. Hence, from Eq. (5.2.6), $\hat{D}_3$ is identically zero, which implies from Eq. $(5.2.6)_2$ that

$$\varphi_{,3} = \frac{\bar{e}_{33}}{\bar{\varepsilon}_{33}} u_{3,3}. \tag{5.2.20}$$

Then

$$T_3 = c_{33}' u_{3,3},$$
$$c_{33}' = \bar{c}_{33} + \bar{e}_{33}^2 / \bar{\varepsilon}_{33}, \tag{5.2.21}$$

where c_{33}' is a piezoelectrically stiffened elastic constant. With Eq. (5.2.20), we write Eq. $(5.2.8)_1$ as an equation for u_3 only, as follows:

$$c'_{33}u_{3,33} = \rho\ddot{u}_3. \tag{5.2.22}$$

The eigenvalue problem for free vibrations of the rod with mechanically free end conditions is

$$c'_{33}u_{3,33} + \rho\omega^2 u_3 = 0, \quad -L/2 < x_3 < L/2,$$
$$c'_{33}u_{3,3} = 0, \quad x_3 = \pm L/2. \tag{5.2.23}$$

The solution of $\omega = 0$ and $u_3 =$ constant represents a rigid-body mode. For the rest of the modes, we try $u_3 = \sin kx_1$. Then, from Eq. (5.2.23)$_1$,

$$k = \omega\sqrt{\frac{\rho}{c'_{33}}}. \tag{5.2.24}$$

To satisfy Eq. (5.2.23)$_2$, we must have

$$\cos k\frac{L}{2} = 0, \quad \Rightarrow \quad k^{(n)}\frac{L}{2} = \frac{n\pi}{2}, \quad n = 1, 3, 5, \ldots, \tag{5.2.25}$$

or

$$\omega^{(n)}\sqrt{\frac{\rho}{c'_{33}}}\frac{L}{2} = \frac{n\pi}{2}, \quad \omega^{(n)} = \frac{n\pi}{L}\sqrt{\frac{c'_{33}}{\rho}}, \quad n = 1, 3, 5, \ldots. \tag{5.2.26}$$

Similarly, by considering $u_3 = \cos kx_1$, the following frequencies can be determined:

$$\omega^{(n)} = \frac{n\pi}{L}\sqrt{\frac{c'_{33}}{\rho}}, \quad n = 2, 4, 6, \ldots. \tag{5.2.27}$$

Equations (5.2.26) and (5.2.27) show that $\bar{e}_{33}$ raises the resonance frequencies through c'_{33}, which shows the piezoelectric stiffening effect.

5.3 Composite Piezoelectric Rods with Transverse Poling

Consider the longitudinal extension of a composite rod consisting of an elastic dielectric layer and two identical piezoelectric dielectric

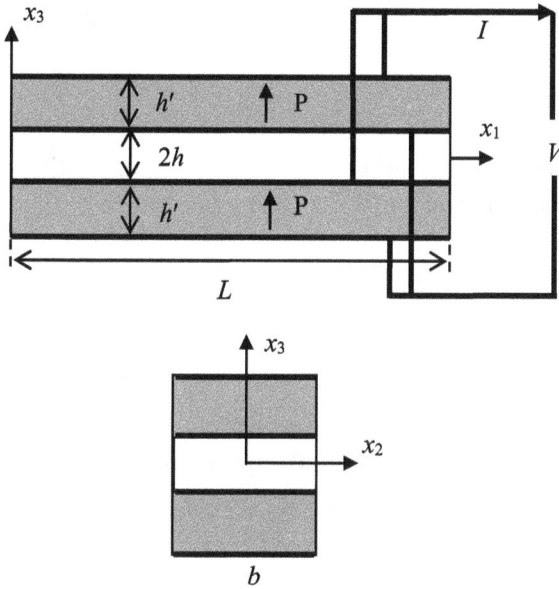

Fig. 5.7. A composite rod with transverse poling and its cross-section.

layers of ceramics poled along x_3 (see Fig. 5.7) [34]. We assume $L \gg h$ and $L \gg b \gg h'$.

For thin rods, we make the following approximations throughout the composite rod (stress relaxation):

$$T_1 = T_1(x_1, t),$$
$$T_2 = T_3 = T_4 = T_5 = T_6 = 0. \tag{5.3.1}$$

The extensional displacement is approximated by

$$u_1 = u_1(x_1, t). \tag{5.3.2}$$

The corresponding axial strain is

$$S_1 = u_{1,1}. \tag{5.3.3}$$

Let the electric potential at the bottom electrodes of the two piezoelectric layers be zero and that at the top electrodes be V. Then the electric fields in the piezoelectric layers are approximated by

$$E_1 = 0, \quad E_2 = 0, \quad E_3 = -\frac{V}{h'}. \tag{5.3.4}$$

The stress–strain relation of the isotropic elastic layer in the middle is

$$T_1 = ES_1, \tag{5.3.5}$$

where E is Young's modulus. The relevant constitutive relations for the piezoelectric layers can be written as

$$S_1 = s_{11}T_1 + d_{31}E_3,$$
$$D_3 = d_{31}T_1 + \varepsilon_{33}E_3. \tag{5.3.6}$$

From Eq. (5.3.6), we solve for the axial stress T_1 and the transverse electric displacement D_3

$$T_1 = \bar{c}_{11}S_1 - \bar{e}_{31}E_3,$$
$$D_3 = \bar{e}_{31}S_1 + \bar{\varepsilon}_{33}E_3, \tag{5.3.7}$$

where

$$\bar{c}_{11} = s_{11}^{-1}, \quad \bar{e}_{31} = s_{11}^{-1}d_{31},$$
$$\bar{\varepsilon}_{33} = \varepsilon_{33}(1 - k_{31}^2), \quad k_{31}^2 = d_{31}^2/(\varepsilon_{33}s_{11}). \tag{5.3.8}$$

The total extensional force N is obtained by integrating the axial stress over the entire cross-section:

$$N = \int_A T_1 \, dx_2 \, dx_3 = \hat{c}S_1 - \hat{e}E_3, \tag{5.3.9}$$

where

$$\hat{c} = 2hbE + 2h'b\bar{c}_{11},$$
$$\hat{e} = 2h'b\bar{e}_{31}. \tag{5.3.10}$$

$\hat{c}$ is the extensional rigidity of the composite rod. The equation of motion is obtained by applying Newton's second law to the differential element in Fig. 5.8 in the axial direction. We have

$$(N + dN) - N + F_1 \, dx_1 = (\rho 2hb + \rho' 2h'b)(dx_1)\ddot{u}_1, \tag{5.3.11}$$

or

$$\frac{\partial N}{\partial x_1} + F_1 = 2b(\rho h + \rho' h')\ddot{u}_1, \tag{5.3.12}$$

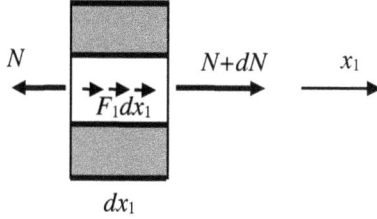

Fig. 5.8.   A differential element of the rod.

where ρ and ρ' are the mass densities of the elastic and piezoelectric layers. F_1 is the mechanical load in the x_1 direction per unit length of the rod.

The free charge on the top electrode at $x_3 = h + h'$ is given by

$$Q^e = b \int_0^L (-D_3) dx_1. \tag{5.3.13}$$

The current flowing out of the top electrodes of the two piezoelectric layers together is

$$I = -2\dot{Q}^e. \tag{5.3.14}$$

When the motion is time harmonic, using the usual complex notation, we write

$$
\begin{aligned}
I &= \text{Re}\{\bar{I}\exp(i\omega t)\}, \\
V &= \text{Re}\{\bar{V}\exp(i\omega t)\}.
\end{aligned}
\tag{5.3.15}
$$

The electrodes may be connected to a circuit whose impedance is Z. Then we have the following circuit equation:

$$\bar{I} = \bar{V}/Z. \tag{5.3.16}$$

As a simple example of the actuation of the rod by a known voltage V, consider a static and mechanically free rod with $F_1 = 0$. In this case, $N = 0$. From Eqs. (5.3.9), (5.3.4) and (5.3.10), we obtain the axial strain as

$$S_1 = \frac{\hat{e}}{\hat{c}} E_3 = -\frac{h'b\bar{e}_{31}}{hbE + h'b\bar{c}_{11}} \frac{V}{h'}. \tag{5.3.17}$$

Conversely, for a simple example of the sensing of an axial strain S_1 in the rod produced by some mechanical load, consider a rod in static and uniform extension with an open circuit between the top and bottom electrodes ($Z = \infty$). Then $Q^e = 0$ and $D_3 = 0$. From Eqs. (5.3.7)$_2$ and (5.3.4), we obtain the output voltage as

$$V = \frac{\bar{e}_{31}}{\bar{\varepsilon}_{33}} h' S_1. \qquad (5.3.18)$$

5.4 Composite Piezoelectric Rods with Axial Poling

Consider the composite rod shown in Fig. 5.9 [11]. It consists of two identical piezoelectric ceramic dielectric layers with axial poling and a nonpiezoelectric elastic dielectric layer in the middle. Its length is much larger than the characteristic dimension of the cross-section. (x, y, z) corresponding to $(1, 2, 3)$.

The rod is in axial extension along x_3 or z. We denote the relevant axial fields by

$$\begin{aligned} u_3 = u, \quad S_3 = S, \quad T_3 = T, \\ E_3 = E, \quad D_3 = D. \end{aligned} \qquad (5.4.1)$$

For a one-dimensional model, we have, approximately,

$$u \cong u(z, t), \quad \varphi \cong \varphi(z, t). \qquad (5.4.2)$$

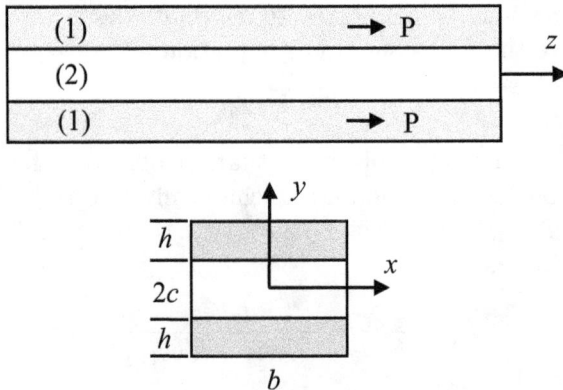

Fig. 5.9.   A composite rod with axial poling and its cross-section.

The relevant strain–displacement relation and electric field–potential relation are

$$S = \frac{\partial u}{\partial z}, \quad E = -\frac{\partial \varphi}{\partial z}. \tag{5.4.3}$$

For the piezoelectric layers, the relevant constitutive relations are

$$\begin{aligned} S &= s_{33}^E T + d_{33} E, \\ D &= d_{33} T + \varepsilon_{33}^T E, \end{aligned} \tag{5.4.4}$$

where $T_1 = T_2 = 0$ has been used. Equation (5.4.4) can be rewritten as

$$\begin{aligned} T &= \bar{c}^{(1)} S - \bar{e}^{(1)} E, \\ D &= \bar{e}^{(1)} S + \bar{\varepsilon}^{(1)} E, \end{aligned} \tag{5.4.5}$$

where $\bar{c}^{(1)}$, $\bar{e}^{(1)}$ and $\bar{\varepsilon}^{(1)}$ are the effective one-dimensional elastic, piezoelectric and dielectric constants. They are related to the three-dimensional material constants through

$$\begin{aligned} \bar{c}^{(1)} &= 1/s_{33}^{E(1)}, \quad \bar{e}^{(1)} = d_{33}^{(1)}/s_{33}^{E(1)}, \\ \bar{\varepsilon}^{(1)} &= \varepsilon_{33}^{T(1)} - (d_{33}^{(1)})^2/s_{33}^{E(1)}. \end{aligned} \tag{5.4.6}$$

Similarly, for the elastic layer,

$$\begin{aligned} T &= \bar{c}^{(2)} S, \\ D &= \bar{\varepsilon}^{(2)} E, \end{aligned} \tag{5.4.7}$$

where

$$\bar{c}^{(2)} = 1/s_{33}^{(2)}, \quad \bar{\varepsilon}^{(2)} = \varepsilon_{33}^{(2)}. \tag{5.4.8}$$

For the composite rod, the total axial force is given by

$$\begin{aligned} N &= \left(\bar{c}^{(1)} S - \bar{e}^{(1)} E \right) A^{(1)} + \left(\bar{c}^{(2)} S \right) A^{(2)} \\ &= \left(\bar{c}^{(1)} A^{(1)} + \bar{c}^{(2)} A^{(2)} \right) S - \bar{e}^{(1)} A^{(1)} E \\ &= \hat{c} S - \hat{e} E = \hat{c} \frac{\partial u}{\partial z} + \hat{e} \frac{\partial \varphi}{\partial z}, \end{aligned} \tag{5.4.9}$$

where Eq. (5.4.3) has been used and

$$\hat{c} = \bar{c}^{(1)} A^{(1)} + \bar{c}^{(2)} A^{(2)}, \quad \hat{e} = \bar{e}^{(1)} A^{(1)},$$
$$A^{(1)} = 2bh, \quad A^{(2)} = 2bc. \tag{5.4.10}$$

$A^{(1)}$ and $A^{(2)}$ are the cross-sectional areas of the piezoelectric and elastic layers, respectively. The total axial electric displacement over the cross-section is

$$\hat{D} = \left(\bar{e}^{(1)} S + \bar{\varepsilon}^{(1)} E\right) A^{(1)} + \left(\bar{\varepsilon}^{(2)} E\right) A^{(2)}$$

$$= \bar{e}^{(1)} A^{(1)} S + \left(\bar{\varepsilon}^{(1)} A^{(1)} + \bar{\varepsilon}^{(2)} A^{(2)}\right) E \tag{5.4.11}$$

$$= \hat{e} S + \hat{\varepsilon} E = \hat{e} \frac{\partial u}{\partial z} - \hat{\varepsilon} \frac{\partial \varphi}{\partial z},$$

where

$$\hat{\varepsilon} = \bar{\varepsilon}^{(1)} A^{(1)} + \bar{\varepsilon}^{(2)} A^{(2)}. \tag{5.4.12}$$

The equation of motion in the axial direction can be obtained as follows by applying Newton's second law to the differential element of the composite rod with length dz shown in Fig. 5.10:

$$\frac{\partial N}{\partial z} + F_z = 2b \left(\rho^{(1)} h + \rho^{(2)} c\right) \ddot{u}, \tag{5.4.13}$$

where $F(z,t)$ is the distributed axial mechanical load per unit length of the rod. Similarly (see Fig. 5.5), the charge equation of electrostatics is

$$\frac{\partial \hat{D}}{\partial z} = 0. \tag{5.4.14}$$

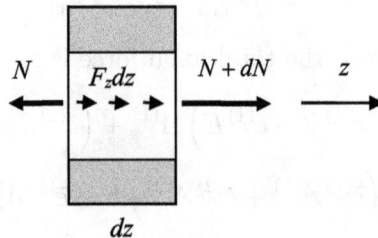

Fig. 5.10. A differential element of the composite rod.

With substitutions from Eqs. (5.4.9) and (5.4.11), we can write Eqs. (5.4.13) and (5.4.14) as two equations for u and φ.

5.5 Piezoelectric Semiconductor Rods

Consider extensional motions of the piezoelectric semiconductor rod of hexagonal crystals such as ZnO shown in Fig. 5.11 [36]. The c-axis of the crystal is along x_3. The cross-sectional area is A.

For extension of thin rods, the fields of interest are approximately uniform over a cross-section:

$$u_3 \cong u_3(x_3, t), \quad \varphi \cong \varphi(x_3, t),$$
$$\Delta p \cong \Delta p(x_3, t), \quad \Delta n \cong \Delta n(x_3, t). \tag{5.5.1}$$

The axial strain and electric field are

$$S_3 = \frac{\partial u_3}{\partial x_3}, \quad E_3 = -\frac{\partial \varphi}{\partial x_3}. \tag{5.5.2}$$

For the piezoelectric constitutive relations, we begin with

$$S_3 = s_{33}T_3 + d_{33}E_3,$$
$$D_3 = d_{33}T_3 + \varepsilon_{33}E_3, \tag{5.5.3}$$

where $T_1 = T_2 \cong 0$ for thin rods has been assumed. Equation (5.5.3) can be rewritten in the following form:

$$T_3 = \bar{c}_{33}S_3 - \bar{e}_{33}E_3 = \bar{c}_{33}\frac{\partial u_3}{\partial x_3} + \bar{e}_{33}\frac{\partial \varphi}{\partial x_3},$$
$$D_3 = \bar{e}_{33}S_3 + \bar{\varepsilon}_{33}E_3 = \bar{e}_{33}\frac{\partial u_3}{\partial x_3} - \bar{\varepsilon}_{33}\frac{\partial \varphi}{\partial x_3}, \tag{5.5.4}$$

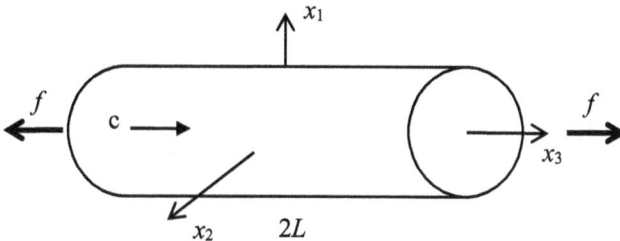

Fig. 5.11. A piezoelectric semiconductor rod of hexagonal crystals.

where Eq. (5.5.2) has been used and the effective one-dimensional material constants are defined by

$$\bar{c}_{33} = 1/s_{33}^E, \quad \bar{e}_{33} = d_{33}/s_{33}^E, \quad \bar{\varepsilon}_{33} = \varepsilon_{33}^T - d_{33}^2/s_{33}^E. \quad (5.5.5)$$

The axial constitutive relations for the current densities are directly reduced from the three-dimensional ones as

$$J_3^p = qp_0\mu_{33}^p E_3 - qD_{33}^p \frac{\partial(\Delta p)}{\partial x_3},$$

$$J_3^n = qn_0\mu_{33}^n E_3 + qD_{33}^n \frac{\partial(\Delta n)}{\partial x_3}. \quad (5.5.6)$$

The total axial force N and the total axial electric displacement $\hat{D}_3$ are given by

$$N = T_3 A = \bar{c}_{33}A\frac{\partial u_3}{\partial x_3} + \bar{e}_{33}A\frac{\partial \varphi}{\partial x_3},$$

$$\hat{D}_3 = D_3 A = \bar{e}_{33}A\frac{\partial u_3}{\partial x_3} - \bar{\varepsilon}_{33}A\frac{\partial \varphi}{\partial x_3}. \quad (5.5.7)$$

The equation of motion in the axial direction can be obtained as follows by applying Newton's second law to a differential element of the rod with length dz as shown in Fig. 5.12:

$$\frac{\partial N}{\partial x_3} + F_3 = \rho A\ddot{u}_3, \quad (5.5.8)$$

where $F_3(x_3, t)$ is the distributed axial mechanical load per unit length of the rod.

Similarly (see Fig. 5.5), the charge equation of electrostatics is

$$\frac{\partial \hat{D}_3}{\partial x_3} = Aq(\Delta p - \Delta n). \quad (5.5.9)$$

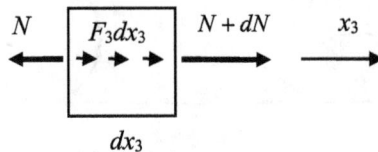

Fig. 5.12. A differential element of the rod under axial mechanical loads.

The continuity equations can be directly reduced from the three-dimensional equations as

$$q\frac{\partial(\Delta p)}{\partial t} = -\frac{\partial J_3^p}{\partial x_3},$$

$$q\frac{\partial(\Delta n)}{\partial t} = \frac{\partial J_3^n}{\partial x_3}. \tag{5.5.10}$$

With substitutions from Eqs. (5.5.7) and (5.5.6), we can write Eqs. (5.5.8)–(5.5.10) as four equations for u_3, φ, Δp and Δn.

As an example, consider an electrically isolated rod within $|x_3| < L$. The rod is under the action of a pair of equal and opposite end forces f which produces an axial stress $T_0 = f/A$. We are interested in the effects of T_0 on the distributions of charge carriers along the rod. There are no concentrated charges at its ends and there are no currents flowing in or out of the rod at its ends. In this case, the boundary conditions are

$$N(\pm L) = f, \quad \hat{D}_3(\pm L) = 0,$$

$$J_3^p(\pm L) = 0, \quad J_3^n(\pm L) = 0. \tag{5.5.11}$$

The rod is assumed to be electrically neutral at the reference state. Therefore, Δp and Δn must satisfy the following charge neutrality conditions:

$$\int_{-L}^{L} \Delta p\, dx = 0, \quad \int_{-L}^{L} \Delta n\, dx = 0. \tag{5.5.12}$$

Equation (5.5.12) is equivalent to one condition because of the charge equation in Eq. (5.5.9) and the electric displacement boundary conditions in Eq. (5.5.11). To determine the mechanical displacement and the electric potential uniquely, we set $x_3 = 0$ as a reference with

$$u_3(0) = 0, \quad \varphi(0) = 0. \tag{5.5.13}$$

The solution can be obtained in a straightforward manner. The electromechanical fields in the rod are found to be

$$u_3 = -\frac{\bar{e}_{33}^2 T_0}{\varepsilon_{33}^T \bar{c}_{33}^2}\frac{\sin h\, kx_3}{k\cos h\, kL} + \frac{T_0}{\bar{c}_{33}}x_3, \tag{5.5.14}$$

$$\varphi = \frac{\bar{e}_{33} T_0}{\varepsilon_{33}^T \bar{c}_{33}}\frac{\sin h\, kx_3}{k\cos h\, kL}, \tag{5.5.15}$$

$$\Delta p = -p_0 \frac{\mu_{33}^p}{D_{33}^p} \frac{\bar{e}_{33} T_0}{\varepsilon_{33}^T \bar{c}_{33}} \frac{\sin h\, kx_3}{k \cos h\, kL},$$

$$\Delta n = n_0 \frac{\mu_{33}^n}{D_{33}^n} \frac{\bar{e}_{33} T_0}{\varepsilon_{33}^T \bar{c}_{33}} \frac{\sin hkx_3}{k \cos h\, kL}, \qquad (5.5.16)$$

where

$$k^2 = \frac{q^2 (p_0 + n_0)}{\varepsilon_{33}^T k_B T}. \qquad (5.5.17)$$

In Eqs. (5.5.14)–(5.5.17), the following has also been used:

$$\varepsilon_{33}^T = \bar{\varepsilon}_{33} \left(1 + \frac{\bar{e}_{33}^2}{\bar{\varepsilon}_{33} \bar{c}_{33}} \right). \qquad (5.5.18)$$

The fields in Eqs. (5.5.14)–(5.5.16) are either symmetric or antisymmetric about the center of the rod. When the rod is a piezoelectric dielectric, these fields are no more than linear functions of x_3. Now they are affected by the mobile charges and are described by hyperbolic functions. It can be verified that when $p_0 \to 0$, $n_0 \to 0$, and hence $k \to 0$, the fields in Eqs. (5.5.14)–(5.5.16) reduce to

$$u_3 = -\frac{\bar{e}_{33}^2 T_0}{\varepsilon_{33}^T \bar{c}_{33}^2} x_3 + \frac{T_0}{\bar{c}_{33}} x_3 = \frac{T_0}{\bar{c}_{33}} x_3 \left(1 - \frac{\bar{e}_{33}^2}{\varepsilon_{33}^T \bar{c}_{33}} \right)$$

$$= s_{33}^E T_0 x_3 \left[1 - \frac{d_{33}^2}{\varepsilon_{33}^T (s_{33}^E)^2} s_{33}^E \right] = s_{33}^E T_0 x_3 \left[1 - \frac{d_{33}^2}{\varepsilon_{33}^T s_{33}^E} \right], \qquad (5.5.19)$$

$$\varphi = \frac{\bar{e}_{33} T_0}{\varepsilon_{33}^T \bar{c}_{33}} x_3 = \frac{d_{33} T_0}{\varepsilon_{33}^T} x_3, \qquad (5.5.20)$$

$$\Delta p = \Delta n = 0. \qquad (5.5.21)$$

Equations (5.5.19) and (5.5.20) are the same as those for the extension of a piezoelectric dielectric rod (see Eq. (5.2.13)). Δp and Δn are antisymmetric about $x_3 = 0$. We calculate the change of mobile charges in half of the rod denoted by Q^e as

$$Q^e = \int_{-L}^{0} \rho^e A \, dx_3 = \int_{-L}^{0} q(\Delta p - \Delta n) A \, dx_3$$

$$= AT_0 \frac{\bar{e}_{33}}{\bar{c}_{33}} \left(1 - \frac{1}{\cos h\, kL} \right), \qquad (5.5.22)$$

or

$$\frac{Q^e}{AT_0} = \frac{Q^e}{f} = \frac{\bar{e}_{33}}{\bar{c}_{33}} \left(1 - \frac{1}{\cos h \, kL} \right) = \gamma. \qquad (5.5.23)$$

γ may be used as a measure of the strength of the coupling of interest, i.e., the effect of f on the redistribution of mobile charges in the rod. When p_0 or n_0 increases, k increases, $\cos h kL$ increases and Q^e increases as expected. The effects of semiconduction appear as a combination through k. For numerical results, consider an n-type ($p \cong 0$) ZnO rod with $2L = 1.2\ \mu$m, cross-sectional area $A = 2.598 \times 10^{-14}\ \text{m}^2$, $n_0 = 10^{21}\ \text{m}^{-3}$ and $f = 8.5\,\text{nN}$ or lower, which produces an axial stress of $T_0 = 3.27 \times 10^5\ \text{N/m}^2$ or lower. For these data, Fig. 5.13 shows the most basic behavior of the rod as a piezoelectric semiconductor, i.e., under the action of the axial force, the electrons in the rod redistribute themselves and assume an axial distribution. Numerical calculation also shows that when n_0 increases, k increases, the hyperbolic functions have larger values and change more rapidly near the ends.

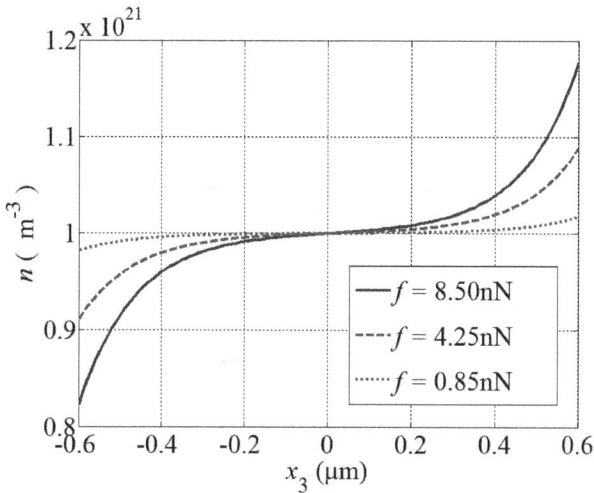

Fig. 5.13.   Axial distribution of electron concentration.

5.6 Composite Thermopiezoelectric Semiconductor Rods

In this section, we study the effects of a known temperature change on the extensional deformation of the composite rod shown in Fig. 5.14 [11,32]. It consists of two identical piezoelectric dielectric layers and a nonpiezoelectric semiconductor layer. Coordinates (x, y, z) correspond to (x_1, x_2, x_3). When the piezoelectric materials are polarized ceramics or hexagonal crystals, the poling direction or the c-axis is along the z axis.

For a thin composite rod, we have, for the entire rod, approximately,

$$u_3 \cong u_3(z, t), \quad \varphi \cong \varphi(z, t). \tag{5.6.1}$$

In the semiconductor layer only, we assume that

$$\Delta p \cong \Delta p(z, t), \quad \Delta n \cong \Delta n(z, t). \tag{5.6.2}$$

Then the axial strain and electric field can be approximated by

$$S_3 = u_{3,3}, \quad E_3 = -\varphi_{,3}, \tag{5.6.3}$$

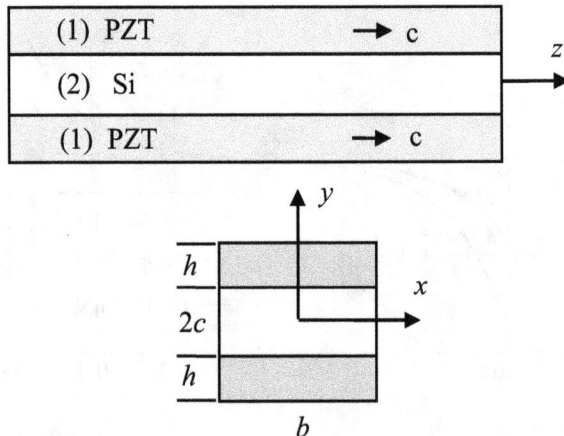

Fig. 5.14. Side view and cross-section of a composite rod.

which are common to all layers. To simplify the notation of the axial fields, we denote

$$u = u_3, \quad S = S_3, \quad T = T_3,$$
$$E = E_3, \quad D = D_3, \quad J^p = J_3^p, \quad J^n = J_3^n. \tag{5.6.4}$$

We also denote the relevant material constants by

$$s = s_{33}^E, \quad d = d_{33}, \quad \varepsilon = \varepsilon_{33}^T, \quad \alpha = \alpha_{33},$$
$$\mu^p = \mu_{33}^p, \quad \mu^n = \mu_{33}^n, \quad D^p = D_{33}^p, \quad D^n = D_{33}^n. \tag{5.6.5}$$

The fields and material constants for the piezoelectric layers will carry a superscript 1 in parentheses, and those of the semiconductor layer, a superscript 2 in parentheses. For the piezoelectric layers, in terms of the elastic compliance, the constitutive relations for the axial strain and electric displacement are

$$S = s^{(1)}T + d^{(1)}E + \alpha^{(1)}\theta,$$
$$D = d^{(1)}T + \varepsilon^{(1)}E + p_3^{(1)}\theta. \tag{5.6.6}$$

From Eq. (5.6.6), we obtain

$$T = \bar{c}^{(1)}S - \bar{e}^{(1)}E - \bar{\lambda}^{(1)}\theta,$$
$$D = \bar{e}^{(1)}S + \bar{\varepsilon}^{(1)}E + \bar{p}^{(1)}\theta, \tag{5.6.7}$$

where

$$\bar{c}^{(1)} = \frac{1}{s^{(1)}}, \quad \bar{e}^{(1)} = \frac{d^{(1)}}{s^{(1)}}, \quad \bar{\lambda}^{(1)} = \frac{\alpha^{(1)}}{s^{(1)}},$$
$$\bar{\varepsilon}^{(1)} = \varepsilon^{(1)} - \frac{(d^{(1)})^2}{s^{(1)}}, \quad \bar{p}^{(1)} = p_3^{(1)} - \frac{d^{(1)}\alpha^{(1)}}{s^{(1)}}. \tag{5.6.8}$$

The semiconductor layer is nonpiezoelectric and does not have pyroelectric coupling. The mechanical and dielectric constitutive relations are

$$S = s^{(2)}T + \alpha^{(2)}\theta, \quad D = \varepsilon^{(2)}E. \tag{5.6.9}$$

Equation (5.6.9) can be rewritten into

$$T = \bar{c}^{(2)} S - \overline{\lambda}^{(2)} \theta, \quad D = \bar{\varepsilon}^{(2)} E, \tag{5.6.10}$$

where

$$\bar{c}^{(2)} = \frac{1}{s^{(2)}}, \quad \overline{\lambda}^{(2)} = \frac{\alpha^{(2)}}{s^{(2)}}, \quad \bar{\varepsilon}^{(2)} = \varepsilon^{(2)}. \tag{5.6.11}$$

The currents exist in the semiconductor layer only and therefore the superscript 2 in parentheses is neglected, as follows:

$$J^p \cong q p_0 \mu^p E - q D^p \frac{d(\Delta p)}{dz},$$
$$J^n \cong q n_0 \mu^n E + q D^n \frac{d(\Delta n)}{dz}. \tag{5.6.12}$$

The total axial force and electric displacement in the composite rod are defined by integrations of T and D over the entire cross-section and are found to be

$$N = \hat{c} S - \hat{e} E - \hat{\lambda} \theta,$$
$$\hat{D} = \hat{e} S + \hat{\varepsilon} E + \hat{p} \theta, \tag{5.6.13}$$

where

$$\hat{c} = \bar{c}^{(1)} A^{(1)} + \bar{c}^{(2)} A^{(2)}, \quad \hat{\lambda} = \overline{\lambda}^{(1)} A^{(1)} + \overline{\lambda}^{(2)} A^{(2)},$$
$$\hat{e} = \bar{e}^{(1)} A^{(1)}, \quad A^{(1)} = 2bh, \quad A^{(2)} = 2bc, \tag{5.6.14}$$
$$\hat{\varepsilon} = \bar{\varepsilon}^{(1)} A^{(1)} + \bar{\varepsilon}^{(2)} A^{(2)}, \quad \hat{p} = \overline{p}^{(1)} A^{(1)}.$$

$A^{(1)}$ and $A^{(2)}$ are the cross-sectional areas of the piezoelectric and semiconductor layers, respectively. The one-dimensional equation of motion is obtained by applying Newton's second law to a differential element of the rod with length dz as shown in Fig. 5.15:

$$\frac{\partial N}{\partial z} + F_z = 2b \left(\rho^{(1)} h + \rho^{(2)} c \right) \ddot{u}, \tag{5.6.15}$$

where $F_z(z,t)$ is the axial load per unit length of the rod.

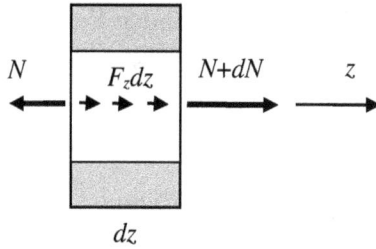

Fig. 5.15.   A differential element of the rod under axial mechanical loads.

Similarly, the one-dimensional charge equation of electrostatics can be obtained as follows by considering the differential element in Fig. 5.15 under electrical loads (see Fig. 5.5):

$$\frac{\partial \hat{D}}{\partial z} = A^{(2)} q (\Delta p - \Delta n). \tag{5.6.16}$$

The one-dimensional continuity equations are simply

$$q \frac{\partial}{\partial t} (\Delta p) = -\frac{\partial J^p}{\partial z},$$
$$q \frac{\partial}{\partial t} (\Delta n) = \frac{\partial J^n}{\partial z}. \tag{5.6.17}$$

With successive substitutions from Eqs. (5.6.13), (5.6.12) and (5.6.3), we can write Eqs. (5.6.15)–(5.6.17) as four equations for u, φ, Δp and Δn.

As an example, consider the static extension of an electrically isolated and mechanically free rod within $|z| < L$. The boundary conditions are

$$N(\pm L) = 0, \quad \hat{D}(\pm L) = 0,$$
$$J^n(\pm L) = 0, \quad J^p(\pm L) = 0. \tag{5.6.18}$$

Δp and Δn satisfy the following global charge neutrality conditions:

$$\int_{-L}^{L} \Delta p \, dz = 0, \quad \int_{-L}^{L} \Delta n \, dz = 0. \tag{5.6.19}$$

Only one of Eq. (5.6.19) is independent. To determine the mechanical displacement and the electric potential uniquely, we set

$$u(0) = 0, \quad \varphi(0) = 0. \tag{5.6.20}$$

Solving Eqs. (5.6.15)–(5.6.17) under Eqs. (5.6.18)–(5.6.20) does not present any mathematical challenge. The fields are found to be

$$u = -\frac{\hat{e}(\hat{e}\hat{\lambda} + \hat{p}\hat{c})\theta}{(\hat{\varepsilon}\hat{c} + \hat{e}^2)\hat{c}} \frac{\sinh kz}{k \cosh kL} + \frac{\hat{\lambda}\theta}{\hat{c}} z, \qquad (5.6.21)$$

$$\varphi = \frac{(\hat{e}\hat{\lambda} + \hat{p}\hat{c})\theta}{(\hat{\varepsilon}\hat{c} + \hat{e}^2)} \frac{\sinh kz}{k \cosh kL}, \qquad (5.6.22)$$

$$\Delta p = \frac{p_0 \mu^p}{D^p} \frac{(\hat{e}\hat{\lambda} + \hat{p}\hat{c})\theta}{(\hat{\varepsilon}\hat{c} + \hat{e}^2)} \frac{\sinh kz}{k \cosh kL}, \qquad (5.6.23)$$

$$\Delta n = -\frac{n_0 \mu^n}{D^n} \frac{(\hat{e}\hat{\lambda} + \hat{p}\hat{c})\theta}{(\hat{\varepsilon}\hat{c} + \hat{e}^2)} \frac{\sinh kz}{k \cosh kL}, \qquad (5.6.24)$$

where

$$k^2 = \left(\frac{p_0 \mu^p}{D^p} + \frac{n_0 \mu^n}{D^n}\right) \frac{q}{\tilde{\varepsilon}}, \quad \tilde{\varepsilon} A^{(2)} = \hat{\varepsilon} + \frac{\hat{e}^2}{\hat{c}}. \qquad (5.6.25)$$

5.7 Piezomagnetic Dielectric Rods under Transverse Fields

This section is on interactions in composite rods of piezomagnetics and piezoelectrics [37]. Specifically, we study the electrical response of a composite rod of a piezoelectric dielectric layer such as polarized ceramics (PZT) between two piezomagnetic (PM) layers under a transverse magnetic field. In Fig. 5.16, the global coordinate system is (x_1, x_2, x_3) or (x, y, z). There is a primed local coordinate system for the PM layers only. The rod is under a transverse magnetic field H along the 3' direction of the PM layers. H is assumed known and uniform in the PM layers. H produces longitudinal extension in the PM layers through the piezomagnetic constant h_{31} in the local coordinate system and thus the extension of the composite rod. The PZT layer is unelectroded on its lateral surfaces. The electric field outside the composite rod is neglected.

In the global coordinate system (x, y, z), we use the following simplified notation for the axial fields:

$$u = u_3, \quad S = S_3, \quad T = T_3, \qquad (5.7.1)$$

$$E = E_3, \quad D = D_3. \qquad (5.7.2)$$

(2) PM ↑ M 3′

(1) PZT → P 1′

3

(2) PM ↑ M

1

2 ←

h

2c

h

b

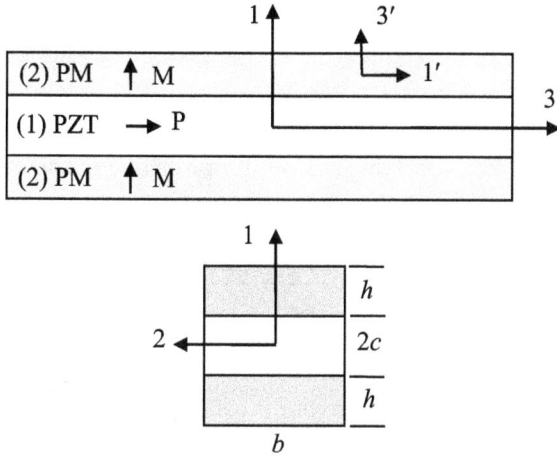

Fig. 5.16. Side view and cross-section of a composite rod.

The following approximate displacement and potential fields are valid throughout the composite rod:

$$u \cong u(z,t), \quad \varphi \cong \varphi(z,t). \tag{5.7.3}$$

Then

$$S = u_{,3}, \quad E = -\varphi_{,3}. \tag{5.7.4}$$

Consider the piezoelectric layer first. Under the stress relaxation that $T_1 = T_2 = 0$, the following effective one-dimensional constitutive relations are the same as those in Eq. (5.4.5):

$$T = \bar{c}^{(1)}S - \bar{e}^{(1)}E,$$
$$D = \bar{e}^{(1)}S + \bar{\varepsilon}^{(1)}E, \tag{5.7.5}$$

where the effective one-dimensional material constants are defined by

$$\bar{c}_{33}^{(1)} = 1/s_{33}^{E(1)}, \quad \bar{e}_{33}^{(1)} = d_{33}^{(1)}/s_{33}^{E(1)},$$
$$\bar{\varepsilon}_{33}^{(1)} = \varepsilon_{33}^{T(1)} - (d_{33}^{(1)})^2/s_{33}^{E(1)}. \tag{5.7.6}$$

For the PM layers, under the stress relaxation for thin rods that in the local coordinate system $T_2' = T_3' \cong 0$, the one-dimensional

constitutive relations for the axial stress and electric displacement
can be written as

$$T = \bar{c}^{(2)} S - \bar{h}^{(2)} H, \quad D = \bar{\varepsilon}^{(2)} E, \tag{5.7.7}$$

where the effective one-dimensional material constants are determined from the stress relaxation as

$$\bar{c}^{(2)} = c_{11}^{(2)} + \frac{2c_{12}^{(2)}(c_{13}^{(2)})^2 - c_{33}^{(2)}(c^{(2)})_{12}^2 - c_{11}^{(2)}(c_{13}^{(2)})^2}{c_{11}^{(2)}c_{33}^{(2)} - (c_{13}^{(2)})^2}, \tag{5.7.8}$$

$$\bar{h}^{(2)} = h_{31}^{(2)} - \frac{c_{12}^{(2)}c_{33}^{(2)}h_{31}^{(2)} - c_{12}^{(2)}c_{13}^{(2)}h_{33}^{(2)} + c_{13}^{(2)}c_{11}^{(2)}h_{33}^{(2)} - (c_{13}^{(2)})^2 h_{31}^{(2)}}{c_{11}^{(2)}c_{33}^{(2)} - (c_{13}^{(2)})^2},$$

$$\tag{5.7.9}$$

$$\bar{\varepsilon}^{(2)} = \varepsilon_{11}^{(2)}. \tag{5.7.10}$$

The total axial force and axial electric displacement are obtained as
follows by integrations over a cross-section of the rod:

$$N = \left(\bar{c}^{(1)} S - \bar{e}^{(1)} E\right) A^{(1)} + \left(\bar{c}^{(2)} S - \bar{h}^{(2)} H\right) A^{(2)}$$
$$= \left(\bar{c}^{(1)} A^{(1)} + \bar{c}^{(2)} A^{(2)}\right) S - \bar{e}^{(1)} A^{(1)} E - \bar{h}^{(2)} A^{(2)} H \quad (5.7.11)$$
$$= \hat{c} S - \hat{e} E - \hat{h} H,$$

$$\hat{D} = \left(\bar{e}^{(1)} S + \bar{\varepsilon}^{(1)} E\right) A^{(1)} + \left(\bar{\varepsilon}^{(2)} E\right) A^{(2)}$$
$$= \bar{e}^{(1)} A^{(1)} S + \left(\bar{\varepsilon}^{(1)} A^{(1)} + \bar{\varepsilon}^{(2)} A^{(2)}\right) E \quad (5.7.12)$$
$$= \hat{e} S + \hat{\varepsilon} E,$$

where

$$\hat{c} = \bar{c}^{(1)} A^{(1)} + \bar{c}^{(2)} A^{(2)}, \quad \hat{e} = \bar{e}^{(1)} A^{(1)}, \quad \hat{h} = \bar{h}^{(2)} A^{(2)},$$
$$\hat{\varepsilon} = \bar{\varepsilon}^{(1)} A^{(1)} + \bar{\varepsilon}^{(2)} A^{(2)}, \quad A^{(1)} = 2bc, \quad A^{(2)} = 2bh. \tag{5.7.13}$$

The equation of motion in the axial direction is obtained by applying
Newton's second law to a differential element of the rod with length
dz as shown in Fig. 5.17:

$$\frac{\partial N}{\partial z} + F_z = 2b\left(\rho^{(1)} c + \rho^{(2)} h\right) \ddot{u}, \tag{5.7.14}$$

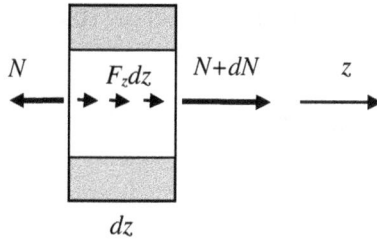

Fig. 5.17. A differential element of the rod under axial mechanical loads.

where $F_z(z,t)$ is the axial mechanical load per unit length of the rod. Similarly, the one-dimensional charge equation of electrostatics is obtained as

$$\frac{\partial \hat{D}}{\partial z} = 0. \tag{5.7.15}$$

With substitutions from Eqs. (5.7.11), (5.7.12) and (5.7.4), we can write Eqs. (5.7.14) and (5.7.15) as two equations for u and φ.

As an example, for the applications of the one-dimensional equations derived, considering the static extension of an electromechanically isolated rod under H only, we have

$$N = \hat{c}S - \hat{e}E - \hat{h}H = 0, \tag{5.7.16}$$

$$\hat{D} = \hat{e}S + \hat{\varepsilon}E = 0. \tag{5.7.17}$$

From Eq. (5.7.17),

$$S = -\frac{\hat{\varepsilon}}{\hat{e}}E. \tag{5.7.18}$$

Substituting Eq. (5.7.18) into Eq. (5.7.16), we have the following effective one-dimensional magnetoelectric effect:

$$E = -\frac{\hat{e}\hat{h}}{\hat{\varepsilon}\hat{c}(1 + \hat{k}^2)}H, \tag{5.7.19}$$

where

$$\hat{k}^2 = \frac{(\hat{e})^2}{\hat{\varepsilon}\hat{c}}. \tag{5.7.20}$$

5.8 Piezomagnetic Semiconductor Rods under Axial Fields

Consider the composite rod shown in Fig. 5.18 [38]. (x, y, z) correspond to (x_1, x_2, x_3). The rod consists of a piezoelectric semiconductor layer labeled by "(1)" such as ZnO and two identical piezomagnetic (PM) layers labeled by "(2)" such as $CoFe_2O_4$. It is under an axial magnetic field H_3 that produces extension of the rod through the piezomagnetic constant h_{33}. The axial semiconduction current does not affect H_3.

We use the following notation for the relevant axial fields and parameters:

$$u = u_3, \quad S = S_3, \quad T = T_3, \tag{5.8.1}$$

$$E = E_3, \quad D = D_3,$$
$$H = H_3, \quad B = B_3, \tag{5.8.2}$$

$$J^p = J_3^p, \quad J^n = J_3^n,$$
$$\mu^p = \mu_{33}^p, \quad D^p = D_{33}^p,$$
$$\mu^n = \mu_{33}^n, \quad D^n = D_{33}^n. \tag{5.8.3}$$

The following approximations of the displacement and potential fields are assumed throughout the rod:

$$u \cong u(z, t), \quad \varphi \cong \varphi(z, t), \quad \psi \cong \psi(z, t). \tag{5.8.4}$$

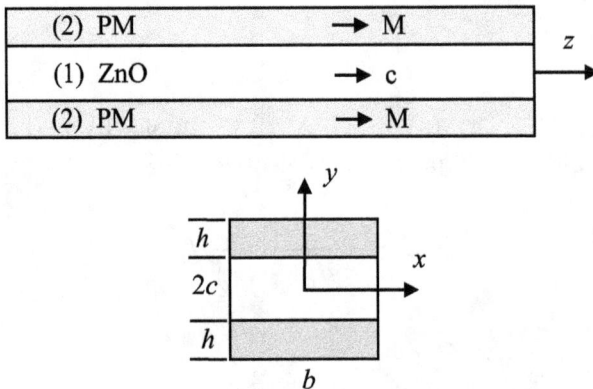

Fig. 5.18.   Side view and cross-section of a composite rod.

Accordingly,

$$S = u_{,3}, \quad E = -\varphi_{,3}, \quad H = -\psi_{,3}. \tag{5.8.5}$$

In Eqs. (5.8.4) and (5.8.5), ψ is a magnetic potential in the one-dimensional sense only. It is simply the integration of the axial magnetic field H along z. In the ZnO layer, we assume that

$$\Delta p \cong \Delta p(z,t), \quad \Delta n \cong \Delta n(z,t). \tag{5.8.6}$$

For constitutive relations, consider the ZnO layer first. We perform the stress relaxation for thin rods using the following relevant constitutive relations:

$$
\begin{aligned}
T_1 &= c_{11}^{(1)} S_1 + c_{12}^{(1)} S_2 + c_{13}^{(1)} S_3 - e_{31}^{(1)} E_3 = 0, \\
T_2 &= c_{12}^{(1)} S_1 + c_{11}^{(1)} S_2 + c_{13}^{(1)} S_3 - e_{31}^{(1)} E_3 = 0, \\
T_3 &= c_{13}^{(1)} S_1 + c_{13}^{(1)} S_2 + c_{33}^{(1)} S_3 - e_{33}^{(3)} E_3,
\end{aligned}
\tag{5.8.7}
$$

$$D_3 = e_{31}^{(1)}(S_1 + S_2) + e_{33}^{(1)} S_3 + \varepsilon_{33}^{(1)} E_3. \tag{5.8.8}$$

We solve Eq. (5.8.7) for expressions of S_1 and S_2 and substitute them into Eq. (5.8.8). This yields the following constitutive relations for the ZnO layer:

$$
\begin{aligned}
T &= \bar{c}^{(1)} S - \bar{e}^{(1)} E, \quad D = \bar{e}^{(1)} S + \bar{\varepsilon}^{(1)} E, \\
B &= \bar{\mu}^{(1)} H,
\end{aligned}
\tag{5.8.9}
$$

where we have added the magnetic constitutive relation and denoted

$$
\begin{aligned}
\bar{c}^{(1)} &= c_{33}^{(1)} - \frac{2(c_{13}^{(1)})^2}{c_{11}^{(1)} + c_{12}^{(1)}}, \quad \bar{e}^{(1)} = e_{33}^{(1)} - \frac{2c_{13}^{(1)} e_{31}^{(1)}}{c_{11}^{(1)} + c_{12}^{(1)}}, \\
\bar{\varepsilon}^{(1)} &= \varepsilon_{33}^{(1)} + \frac{2(e_{31}^{(1)})^2}{c_{11}^{(1)} + c_{12}^{(1)}}, \quad \bar{\mu}^{(1)} = \mu_{33}^{(1)}.
\end{aligned}
\tag{5.8.10}
$$

The linearized constitutive relations for the current densities in the ZnO layer are

$$
\begin{aligned}
J^p &\cong q p_0 \mu^p E - q D^p \frac{\partial(\Delta p)}{\partial z}, \\
J^n &\cong q n_0 \mu^n E + q D^n \frac{\partial(\Delta n)}{\partial z}.
\end{aligned}
\tag{5.8.11}
$$

Similarly, for the PM layers, after the stress relaxation, we obtain

$$T = \bar{c}^{(2)} S - \bar{h}^{(2)} H, \quad D = \bar{\varepsilon}^{(2)} E,$$
$$B = \bar{h}^{(2)} S + \bar{\mu}^{(2)} H, \tag{5.8.12}$$

where

$$\bar{c}^{(2)} = c_{33}^{(2)} - \frac{2(c_{13}^{(2)})^2}{c_{11}^{(2)} + c_{12}^{(2)}}, \quad \bar{h}^{(2)} = h_{33}^{(2)} - \frac{2 c_{13}^{(2)} h_{31}^{(2)}}{c_{11}^{(2)} + c_{12}^{(2)}},$$

$$\bar{\mu}^{(2)} = \mu_{33}^{(2)} + \frac{2(h_{31}^{(2)})^2}{c_{11}^{(2)} + c_{12}^{(2)}}, \quad \bar{\varepsilon}^{(2)} = \varepsilon_{33}^{(2)}. \tag{5.8.13}$$

For the composite rod, the total axial force N over the entire cross-section is calculated from

$$N = \left(\bar{c}^{(1)} S - \bar{e}^{(1)} E \right) A^{(1)} + \left(\bar{c}^{(2)} S - \bar{h}^{(2)} H \right) A^{(2)}$$
$$= \left(\bar{c}^{(1)} A^{(1)} + \bar{c}^{(2)} A^{(2)} \right) S - \bar{e}^{(1)} A^{(1)} E - \bar{h}^{(2)} A^{(2)} H \tag{5.8.14}$$
$$= \hat{c} S - \hat{e} E - \hat{h} H,$$

where

$$\hat{c} = \bar{c}^{(1)} A^{(1)} + \bar{c}^{(2)} A^{(2)}, \quad \hat{e} = \bar{e}^{(1)} A^{(1)}, \quad \hat{h} = \bar{h}^{(2)} A^{(2)},$$
$$A^{(1)} = 2bc, \quad A^{(2)} = 2bh. \tag{5.8.15}$$

Similarly, the total axial electric displacement and magnetic induction are

$$\hat{D} = \left(\bar{e}^{(1)} S + \bar{\varepsilon}^{(1)} E \right) A^{(1)} + \left(\bar{\varepsilon}^{(2)} E \right) A^{(2)}$$
$$= \bar{e}^{(1)} A^{(1)} S + \left(\bar{\varepsilon}^{(1)} A^{(1)} + \bar{\varepsilon}^{(2)} A^{(2)} \right) E = \hat{e} S + \hat{\varepsilon} E, \tag{5.8.16}$$

$$\hat{B} = \left(\bar{\mu}^{(1)} H \right) A^{(1)} + \left(\bar{h}^{(2)} S + \bar{\mu}^{(2)} H \right) A^{(2)}$$
$$= \bar{h}^{(2)} A^{(2)} S + \left(\bar{\mu}^{(1)} A^{(1)} + \bar{\mu}^{(2)} A^{(2)} \right) H = \hat{h} S + \hat{\mu} H, \tag{5.8.17}$$

where

$$\hat{\varepsilon} = \bar{\varepsilon}^{(1)} A^{(1)} + \varepsilon^{(2)} A^{(2)}, \quad \hat{\mu} = \bar{\mu}^{(1)} A^{(1)} + \bar{\mu}^{(2)} A^{(2)}. \tag{5.8.18}$$

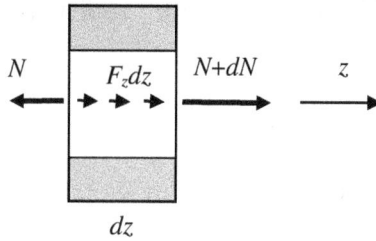

Fig. 5.19. A differential element of the rod under axial mechanical loads.

The equation of motion in the axial direction is obtained as follows by considering a differential element of the rod with length dz as shown in Fig. 5.19:

$$\frac{\partial N}{\partial z} + F_z = 2b \left(\rho^{(1)} c + \rho^{(2)} h \right) \ddot{u}, \qquad (5.8.19)$$

where $F(z, t)$ is the axial mechanical load per unit length of the rod.

Similarly, the one-dimensional charge equation of electrostatics (see Fig. 5.5), the equation for the magnetic induction and the conservation of holes and electrons are

$$\frac{\partial \hat{D}}{\partial z} = q(\Delta p - \Delta n) A^{(1)},$$

$$\frac{\partial \hat{B}}{\partial z} = 0, \qquad (5.8.20)$$

$$q \frac{\partial}{\partial t}(\Delta p) = -\frac{\partial J^p}{\partial z},$$

$$q \frac{\partial}{\partial t}(\Delta n) = \frac{\partial J^n}{\partial z}. \qquad (5.8.21)$$

The substitution of Eqs. (5.8.11), (5.8.14), (5.8.16) and (5.8.17) into Eqs. (5.8.19)–(5.8.21), with the use of Eq. (5.8.5), yields five equations for u, φ, ψ, Δp and Δn.

As an example, consider the static extension of a mechanically free and electrically isolated rod within $-L < z < L$. It is under an axial magnetic field produced by a magnetic potential difference between the two ends. The boundary conditions are

$$N(\pm L) = 0, \quad \hat{D}(\pm L) = 0, \quad \psi(\pm L) = \pm \psi_0,$$
$$J^n(\pm L) = 0, \quad J^p(\pm L) = 0. \tag{5.8.22}$$

Δp and Δn must satisfy the following global charge neutrality conditions:

$$\int_{-L}^{L} \Delta p \, dz = 0, \quad \int_{-L}^{L} \Delta n \, dz = 0. \tag{5.8.23}$$

Only one of Eq. (5.8.23) is independent. The other is implied by integrating Eq. (5.8.20)$_1$ between $-L$ and L and using the boundary conditions on the electric displacement in Eq. (5.8.22), which leads to

$$\int_{-L}^{L} q(\Delta p - \Delta n) dz = 0. \tag{5.8.24}$$

To determine the mechanical displacement and the electric potential uniquely, we set

$$u(0) = 0, \quad \varphi(0) = 0. \tag{5.8.25}$$

Mathematically, we have a system of linear ordinary differential equations with constant coefficients. Its solution can be obtained through a standard procedure. The results are

$$\psi = \frac{\hat{e}^2 \hat{h}^2}{\tilde{c}\hat{c}\tilde{\varepsilon}\hat{\mu}} \frac{\psi_0}{\Delta} \sin \mathrm{h}(kz) + k \cos \mathrm{h}(kL)\frac{\psi_0}{\Delta}z, \tag{5.8.26}$$

$$u_3 = \frac{\hat{e}^2 \hat{h}}{\tilde{c}\hat{c}\tilde{\varepsilon}} \frac{\psi_0}{\Delta} \sin \mathrm{h}(kz) - \frac{\hat{h}k \cos \mathrm{h}(kL)}{\hat{c}} \frac{\psi_0}{\Delta}z, \tag{5.8.27}$$

$$\varphi = -\frac{\hat{e}\hat{h}}{\hat{c}\tilde{\varepsilon}} \frac{\psi_0}{\Delta} \sin \mathrm{h}(kz), \tag{5.8.28}$$

$$\Delta p = \frac{\mu^p p_0}{D^p} \frac{\hat{e}\hat{h}}{\hat{c}\tilde{\varepsilon}} \frac{\psi_0}{\Delta} \sin \mathrm{h}(kz), \tag{5.8.29}$$

$$\Delta n = -\frac{\mu^n n_0}{D^n} \frac{\hat{e}\hat{h}}{\hat{c}\tilde{\varepsilon}} \frac{\psi_0}{\Delta} \sin \mathrm{h}(kz), \tag{5.8.30}$$

where

$$k^2 = \frac{qA^{(1)}}{\tilde{\varepsilon}} \left(\frac{\mu^p}{D^p} p_0 + \frac{\mu^n}{D^n} n_0 \right),$$

$$\tilde{\varepsilon} = \hat{\varepsilon} + \frac{\hat{e}^2}{\tilde{c}}, \quad \tilde{c} = \hat{c} + \frac{\hat{h}^2}{\hat{\mu}}, \tag{5.8.31}$$

$$\Delta = kL \cos h(kL) + \frac{\hat{e}^2 \hat{h}^2}{\tilde{c}\hat{c}\tilde{\varepsilon}\hat{\mu}} \sin h(kL).$$

In the special case of a dielectric rod in static extension under $\psi = \pm\psi_0$ at $z = \pm L$ only, we have $\psi = z\psi_0/L$, $H = -\psi_0/L$ and

$$N = \hat{c}S - \hat{e}E - \hat{h}H = 0,$$

$$\hat{D} = \hat{e}S + \hat{\varepsilon}E = 0. \tag{5.8.32}$$

Eliminating S, we obtain the magnetoelectric coupling as

$$E = -\frac{\hat{e}\hat{h}}{\hat{\varepsilon}\hat{c}(1 + \hat{k}^2)} H, \quad \hat{k}^2 = \frac{(\hat{e})^2}{\hat{\varepsilon}\hat{c}}. \tag{5.8.33}$$

Chapter 6

Torsion of Shafts

This chapter begins with a presentation of some three-dimensional equations in cylindrical coordinates which are convenient in the treatment of torsion of circular shafts. Then the one-dimensional equations for the torsion of a circular elastic shaft in Sec. 1.2 are re-derived using cylindrical coordinates and are generalized to the torsion of circular shafts made from piezoelectric dielectrics or piezoelectric semiconductors. These are followed by a general one-dimensional description of thin rods which is specialized for torsion of rectangular shafts with warping and shear deformation within a cross-section.

6.1 Three-Dimensional Equations in Cylindrical Coordinates

To analyze circular cylindrical structures, it is convenient to use cylindrical coordinates (r, θ, z) defined by

$$x_1 = r \cos \theta, \quad x_2 = r \sin \theta, \quad x_3 = z. \tag{6.1.1}$$

In cylindrical coordinates, we have the following strain–displacement relations:

$$S_{rr} = u_{r,r}, \quad S_{\theta\theta} = \frac{1}{r} u_{\theta,\theta} + \frac{u_r}{r}, \quad S_{zz} = u_{z,z},$$

$$2S_{r\theta} = u_{\theta,r} + \frac{1}{r} u_{r,\theta} - \frac{u_\theta}{r}, \quad 2S_{\theta z} = \frac{1}{r} u_{z,\theta} + u_{\theta,z}, \tag{6.1.2}$$

$$2S_{zr} = u_{r,z} + u_{z,r}.$$

The equations of motion take the following form:

$$\frac{\partial T_{rr}}{\partial r} + \frac{1}{r}\frac{\partial T_{\theta r}}{\partial \theta} + \frac{\partial T_{zr}}{\partial z} + \frac{T_{rr} - T_{\theta\theta}}{r} + f_r = \rho\ddot{u}_r,$$

$$\frac{\partial T_{r\theta}}{\partial r} + \frac{1}{r}\frac{\partial T_{\theta\theta}}{\partial \theta} + \frac{\partial T_{z\theta}}{\partial z} + \frac{2}{r}T_{r\theta} + f_\theta = \rho\ddot{u}_\theta, \qquad (6.1.3)$$

$$\frac{\partial T_{rz}}{\partial r} + \frac{1}{r}\frac{\partial T_{\theta z}}{\partial \theta} + \frac{\partial T_{zz}}{\partial z} + \frac{1}{r}T_{rz} + f_z = \rho\ddot{u}_z.$$

The gradient of a scalar field φ is given by

$$\nabla\varphi = \frac{\partial \varphi}{\partial r}\mathbf{e}_r + \frac{1}{r}\frac{\partial \varphi}{\partial \theta}\mathbf{e}_\theta + \frac{\partial \varphi}{\partial z}\mathbf{e}_z. \qquad (6.1.4)$$

The divergence of a vector field $\mathbf{D}$ is

$$\nabla \cdot \mathbf{D} = \frac{1}{r}(rD_r)_{,r} + \frac{1}{r}D_{\theta,\theta} + D_{z,z}. \qquad (6.1.5)$$

The Laplace operator (Laplacian) on a scalar field φ becomes

$$\nabla^2\varphi = \frac{1}{r}\frac{\partial}{\partial r}\left(r\frac{\partial \varphi}{\partial r}\right) + \frac{1}{r^2}\frac{\partial^2 \varphi}{\partial \theta^2} + \frac{\partial^2 \varphi}{\partial z^2}. \qquad (6.1.6)$$

6.2 Circular Elastic Shafts

Consider the circular cylindrical tube of inner radius a and outer radius b shown in Fig. 6.1. The cross-section is denoted by A. (r, θ, z) correspond to $(2,3,1)$. We re-derive the one-dimensional equations in Sec. 1.2 for torsion from the three-dimensional equations of elasticity in cylindrical coordinates.

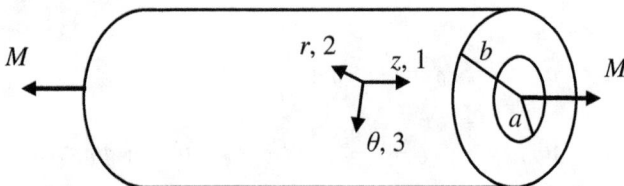

Fig. 6.1.   An elastic circular cylindrical tube in torsion.

Consider the following displacement field:

$$u_1 \cong 0, \quad u_2 \cong 0, \quad u_3 \cong r\psi(z,t). \tag{6.2.1}$$

We limit ourselves to materials without couplings between extensional and shear strains, and couplings among different shear strains. The nontrivial ones of the strain and stress components corresponding to Eq. (6.2.1) are

$$S_5 = 2S_{13} = 2S_{z\theta} = \frac{1}{r}u_{z,\theta} + u_{\theta,z} = r\psi_{,z},$$

$$T_5 = T_{z\theta} = c_{55}r\psi_{,z}. \tag{6.2.2}$$

The stress distribution in Eq. $(6.2.2)_2$ does not produce any net axial extensional force, transverse shear force or bending moment over a cross-section. The only nontrivial resultant over a cross-section is a torque or twisting moment given by

$$M = \int_A T_{z\theta}r dA = \int_a^b T_{z\theta}r(2\pi r dr) = \int_a^b c_{55}r\psi_{,z}2\pi r^2 dr$$

$$= \int_a^b c_{55}\psi_{,z}2\pi r^3 dr = c_{55}I_p\psi_{,z}, \tag{6.2.3}$$

where

$$I_p = \frac{\pi}{2}(b^4 - a^4). \tag{6.2.4}$$

I_p is the polar moment of inertia of the cross-section about its center. The equation for torsional motion is obtained by the moment equation of the differential element in Fig. 6.2:

$$M_{,z} + m_z = \rho I_p\ddot{\psi}, \tag{6.2.5}$$

where $m_z(z,t)$ is the distributed torsional load per unit length of the shaft.

Substituting Eq. (6.2.3) into Eq. (6.2.5), we obtain a single equation for ψ as follows:

$$(c_{55}I_p\psi_{,z})_{,z} + m_z = \rho I_p\ddot{\psi}. \tag{6.2.6}$$

For a uniform shaft with constant $c_{55}I_p$, Eq. (6.2.6) reduces to

$$c_{55}I_p\psi_{,zz} + m_z = \rho I_p\ddot{\psi}. \tag{6.2.7}$$

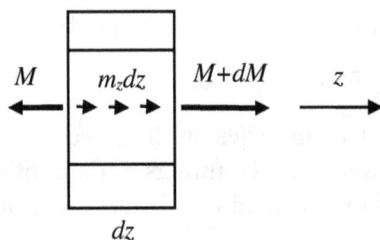

Fig. 6.2.   A differential element of the shaft.

Equation (6.2.5) can also be obtained from the corresponding three-dimensional equation of motion without using Fig. 6.2, which is not the main interest of this book and is not pursued.

As an example, consider waves propagating in a free and unbounded shaft described by

$$\psi = \Psi \exp[i(kx - \omega t)]. \tag{6.2.8}$$

The substitution of Eq. (6.2.8) into Eq. (6.2.7) yields

$$-c_{55}\Psi k^2 = -\rho\Psi\omega^2, \tag{6.2.9}$$

which determines the torsional wave speed as

$$\frac{\omega}{k} = \sqrt{\frac{c_{55}}{\rho}}. \tag{6.2.10}$$

In the case of isotropic materials, $c_{55} = G$, the shear modulus. Then Eq. (6.2.10) reduces to Eq. (1.2.13).

6.3   Circular Piezoelectric Dielectric Shafts

Consider the circular cylindrical tube of inner radius a and outer radius b shown in Fig. 6.3. The cross-section is denoted by A. The cylinder is made of polarized ceramics with circumferential or tangential poling along θ. We choose (r, θ, z) to correspond to $(2,3,1)$ so that the poling direction corresponds to 3. The lateral cylindrical surfaces of the cylinder are traction free and are unelectroded. The free space electric field is neglected as usual. We proceed to develop a one-dimensional model for the torsion of the cylinder [39]. The following derivation is slightly different from that in [39].

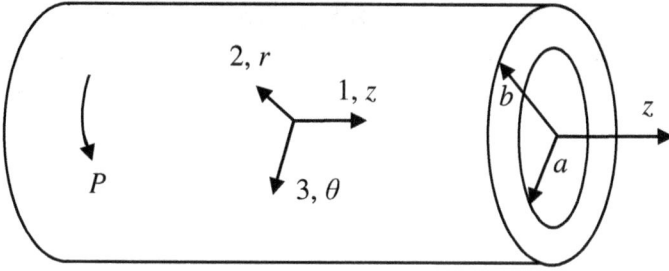

Fig. 6.3. A piezoelectric dielectric circular cylindrical tube.

Consider the following displacement and potential fields:

$$u_1 \cong 0, \quad u_2 \cong 0, \quad u_3 \cong r\psi(z,t),$$
$$\varphi \cong \varphi(z,t). \tag{6.3.1}$$

The nontrivial components of the strain, electric field, stress and electric displacement fields are

$$S_5 = 2S_{\theta z} = r\psi_{,z}, \quad E_1 = E_z = -\varphi_{,z}, \tag{6.3.2}$$

$$T_5 = T_{\theta z} = c_{55}r\psi_{,z} + e_{15}\varphi_{,z},$$
$$D_1 = D_z = e_{15}r\psi_{,z} - \varepsilon_{11}\varphi_{,z}. \tag{6.3.3}$$

The stress distribution in Eq. $(6.3.3)_1$ does not produce any net extensional or shear forces and bending moments over a cross-section. The torque at a cross-section is given by

$$
\begin{aligned}
M &= \int_A T_{z\theta} r \, dA = \int_a^b T_{z\theta} r (2\pi r \, dr) \\
&= \int_a^b (c_{55}r\psi_{,z} + e_{15}\varphi_{,z}) 2\pi r^2 \, dr \\
&= \int_a^b (c_{55}\psi_{,z} 2\pi r^3 + e_{15}\varphi_{,z} 2\pi r^2) dr \\
&= c_{55}I_p\psi_{,z} + e_{15}B\varphi_{,z},
\end{aligned}
\tag{6.3.4}
$$

where

$$I_p = \frac{\pi}{2}(b^4 - a^4), \quad B = \frac{2\pi}{3}(b^3 - a^3). \tag{6.3.5}$$

I_p is the polar moment of inertia of the cross-section about its center. The total axial electric displacement over a cross-section is calculated from

$$\hat{D}_z = \int_A D_z dA = \int_a^b D_z 2\pi r\, dr$$

$$= \int_a^b (e_{15} r\psi_{,z} - \varepsilon_{11}\varphi_{,z})2\pi r\, dr \qquad (6.3.6)$$

$$= \int_a^b (e_{15}\psi_{,z}2\pi r^2 - \varepsilon_{11}\varphi_{,z}2\pi r)dr$$

$$= e_{15} B\psi_{,z} - \varepsilon_{11} A\varphi_{,z},$$

where

$$A = \pi(b^2 - a^2). \qquad (6.3.7)$$

The equation for the torsional motion is obtained by the moment equation of the differential element in Fig. 6.4:

$$M_{,z} + m_z = \rho I_p \ddot{\psi}, \qquad (6.3.8)$$

where $m(z,t)$ is the distributed twisting moment per unit length of the shaft.

Similarly, the one-dimensional charge equation of electrostatics can be obtained from the differential element in Fig. 6.5:

$$\hat{D}_{z,z} = 0. \qquad (6.3.9)$$

Substituting Eqs. (6.3.4) and (6.3.6) into Eqs. (6.3.8) and (6.3.9), we arrive at two equations for ψ and φ as

$$(c_{55} I_p \psi_{,z} + e_{15} B\varphi_{,z})_{,z} + m_z = \rho I_p \ddot{\psi},$$

$$(e_{15} B\psi_{,z} - \varepsilon_{11} A\varphi_{,z})_{,z} = 0. \qquad (6.3.10)$$

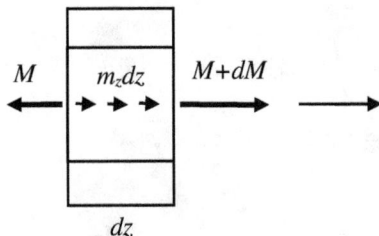

Fig. 6.4.   A differential element of the shaft under mechanical loads.

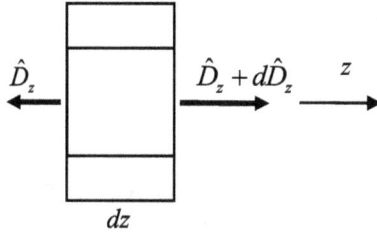

Fig. 6.5. A differential element of the shaft under electrical loads.

For a uniform shaft, Eq. (6.3.10) reduces to

$$c_{55}I_p\psi_{,zz} + e_{15}B\varphi_{,zz} + m_z = \rho I_p\ddot{\psi},$$
$$e_{15}B\psi_{,zz} - \varepsilon_{11}A\varphi_{,zz} = 0. \tag{6.3.11}$$

As an example, consider the propagation of the following waves:

$$\begin{bmatrix} \psi \\ \varphi \end{bmatrix} = \begin{bmatrix} \Psi \\ \Phi \end{bmatrix} \exp[i(kz - \omega t)]. \tag{6.3.12}$$

The substitution of Eq. (6.3.12) into Eq. (6.3.11) results in

$$c_{55}I_p k^2\Psi + e_{15}Bk^2\Phi = \rho I_p\omega^2\Psi,$$
$$e_{15}Bk^2\Psi - \varepsilon_{11}Ak^2\Phi = 0. \tag{6.3.13}$$

From Eq. (6.3.13)$_2$,

$$\Phi = \frac{e_{15}B}{\varepsilon_{11}A}\Psi. \tag{6.3.14}$$

Substituting Eq. (6.3.14) into Eq. (6.3.13)$_1$, we obtain the wave speed as

$$\frac{\omega}{k} = \sqrt{\frac{c'_{55}}{\rho}}, \tag{6.3.15}$$

where c'_{55} is a piezoelectrically stiffened elastic constant:

$$c'_{55} = c_{55}\left(1 + k_{15}^2\frac{B^2}{I_pA}\right), \quad k_{15}^2 = \frac{e_{15}^2}{\varepsilon_{11}c_{55}}. \tag{6.3.16}$$

6.4 Circular Piezoelectric Semiconductor Shafts

Consider the composite core-shell shaft with a circular cross-section shown in Fig. 6.6 [40]. The cylindrical coordinates (r, θ, z) correspond to the tensor indices of $(2,3,1)$. The core within $r < a$ is a nonpiezoelectric semiconductor such as silicon which is a cubic crystal and is indicated as material (1). The outer shell in $a < r < b$ is a piezoelectric dielectric such as polarized ceramics with circumferential or tangential poling, which is labeled as material (2).

The displacement field for torsion is approximated by the angle of twist ψ through

$$u_1 \cong 0, \quad u_2 \cong 0, \quad u_3 \cong r\psi(z,t). \tag{6.4.1}$$

For a thin shaft, the electric potential may be approximated by

$$\varphi \cong \varphi(z,t). \tag{6.4.2}$$

Accordingly, the relevant components of the strain and electric fields are

$$S_5 = 2S_{\theta z} = u_{\theta,z} = u_{3,z} = r\psi_{,z},$$
$$E_1 = E_z = -\varphi_{,z}. \tag{6.4.3}$$

Equations (6.4.1)–(6.4.3) are approximately valid for the entire shaft. For the charge carriers in the semiconductor core, we make the following approximation:

$$\Delta p \cong \Delta p(z,t), \quad \Delta n \cong \Delta n(z,t). \tag{6.4.4}$$

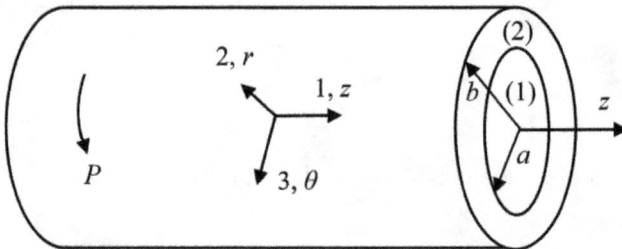

Fig. 6.6. A composite piezoelectric semiconductor circular shaft.

In the semiconductor core, the relevant constitutive relations take the following form:

$$T_5 = T_{\theta z} = c_{55}^{(1)} S_5 = c_{55}^{(1)} r \psi_{,z},$$

$$D_1 = D_z = \varepsilon_{11}^{(1)} E_1 = -\varepsilon_{11}^{(1)} \varphi_{,z},$$

$$\qquad\qquad (6.4.5)$$

$$J_1^p = J_z^p = q p_0 \mu_{11}^p E_1 - q D_{11}^p (\Delta p)_{,1}$$

$$= -q p_0 \mu_{11}^p \varphi_{,z} - q D_{11}^p (\Delta p)_{,z},$$

$$J_1^n = J_z^n = q n_0 \mu_{11}^n E_1 + q D_{11}^n (\Delta n)_{,1}$$

$$\qquad\qquad (6.4.6)$$

$$= -q n_0 \mu_{11}^n \varphi_{,z} + q D_{11}^n (\Delta n)_{,z}.$$

In the piezoelectric dielectric shell where there are no currents, the relevant constitutive relations are

$$T_5 = T_{\theta z} = c_{55}^{(2)} S_5 - e_{15}^{(2)} E_1 = c_{55}^{(2)} r \psi_{,z} + e_{15}^{(2)} \varphi_{,z},$$

$$D_1 = D_z = e_{15}^{(2)} S_5 + \varepsilon_{11}^{(2)} E_1 = e_{15}^{(2)} r \psi_{,z} - \varepsilon_{11}^{(2)} \varphi_{,z}.$$

$$\qquad\qquad (6.4.7)$$

The total torque or twisting moment is given by the following integration over the cross-sectional area A:

$$M = \int_A T_{z\theta} r \, dA = \hat{c} \psi_{,z} + \hat{e} \varphi_{,z}, \qquad\qquad (6.4.8)$$

where we have denoted

$$\hat{c} = 2\pi c_{55}^{(1)} \frac{1}{4} a^4 + 2\pi c_{55}^{(2)} \frac{1}{4} (b^4 - a^4),$$

$$\hat{e} = 2\pi e_{15}^{(2)} \frac{1}{3} (b^3 - a^3).$$

$$\qquad\qquad (6.4.9)$$

The total axial electric displacement over a cross-section is calculated from

$$\hat{D} = \int_A D_z \, dA = \hat{e} \psi_{,z} - \hat{\varepsilon} \varphi_{,z}, \qquad\qquad (6.4.10)$$

where

$$\hat{\varepsilon} = 2\pi \varepsilon_{11}^{(1)} \frac{1}{2} a^2 + 2\pi \varepsilon_{11}^{(2)} \frac{1}{2} (b^2 - a^2). \qquad\qquad (6.4.11)$$

The equation for the torsional motion is obtained by considering a differential element of the shaft with length dz as shown in Fig. 6.7.

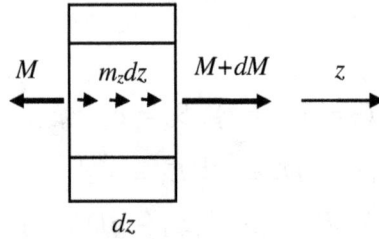

Fig. 6.7. A differential element of the shaft.

Taking moment about the z axis, we obtain

$$(M + dM) - M + m_z\, dz = dz \int_A \rho \frac{\partial^2 u_3}{\partial t^2} r\, dA$$

$$= dz \int_0^a \rho^{(1)} r \frac{\partial^2 \psi}{\partial t^2} r(2\pi r) dr + dz \int_a^b \rho^{(2)} r \frac{\partial^2 \psi}{\partial t^2} r(2\pi r) dr \quad (6.4.12)$$

$$= 2\pi \rho^{(1)} \frac{\partial^2 \psi}{\partial t^2} \frac{a^4}{4} dz + 2\pi \rho^{(2)} \frac{\partial^2 \psi}{\partial t^2} \frac{1}{4}(b^4 - a^4) dz = \bar{I}_p \frac{\partial^2 \psi}{\partial t^2} dz,$$

where

$$\bar{I}_p = 2\pi \rho^{(1)} \frac{a^4}{4} + 2\pi \rho^{(2)} \frac{1}{4}(b^4 - a^4). \quad (6.4.13)$$

Equation (6.4.12) can be written as

$$\frac{\partial M}{\partial z} + m_z = \bar{I}_p \frac{\partial^2 \psi}{\partial t^2}. \quad (6.4.14)$$

Similarly, the one-dimensional charge equation of electrostatics is obtained by applying electrical loads to the element in Fig. 6.7 (see Fig. 6.5), which results in

$$\frac{\partial \hat{D}}{\partial z} = q(\Delta p - \Delta n)\pi a^2. \quad (6.4.15)$$

The one-dimensional conservation of holes and electrons can be directly reduced from the corresponding three-dimensional ones as

$$q\frac{\partial}{\partial t}(\Delta p) = -\frac{\partial J_z^p}{\partial z},$$

$$q\frac{\partial}{\partial t}(\Delta n) = \frac{\partial J_z^n}{\partial z}. \quad (6.4.16)$$

When Eqs. (6.4.8), (6.4.10) and (6.4.6) are substituted into Eqs. (6.4.14)–(6.4.16), it results in four equations for ψ, φ, Δp and Δn.

As an example, consider the static torsion of a finite shaft within $(-L, L)$ under a constant end torque M_0. We limit ourselves to an n-type semiconductor with $p_0 \cong 0$ and $\Delta p \cong 0$. The shaft is electrically open at its ends. The boundary conditions are

$$M(\pm L) = M_0, \quad \hat{D}(\pm L) = 0, \quad J_z^n(\pm L) = 0. \tag{6.4.17}$$

To fix the rigid-body rotation and to make the electric potential unique, we set

$$\psi(0) = 0, \quad \varphi(0) = 0 \tag{6.4.18}$$

as a reference. For static torsion, Eqs. (6.4.14)–(6.4.16) lead to linear ordinary differential equations with constant coefficients. The solution can be obtained in a systematic manner. The results are

$$\psi(z) = \frac{M_0}{\hat{c}} z - \frac{\hat{e}^2 M_0}{\hat{c}^2 \tilde{\varepsilon} \pi a^2} \frac{\sin h(kz)}{k \cos h(kL)}, \tag{6.4.19}$$

$$\varphi(z) = \frac{\hat{e} M_0}{\hat{c} \tilde{\varepsilon} \pi a^2} \frac{\sin h(kz)}{k \cos h(kL)}, \tag{6.4.20}$$

$$\Delta n = \frac{\hat{e} k M_0}{q \hat{c} \pi a^2} \frac{\sin h(kz)}{\cos h(kL)}, \tag{6.4.21}$$

where

$$k^2 = \frac{n_0 \mu_{11}^n}{D_{11}^n} \cdot \frac{q}{\tilde{\varepsilon}} = \frac{1}{\lambda_D^2}, \quad \tilde{\varepsilon} = \frac{\hat{e}^2 + \hat{c}\tilde{\varepsilon}}{\pi a^2 \hat{c}}. \tag{6.4.22}$$

We calculate the net charge in the right half of the shaft through

$$Q^e = \int_0^L q(-\Delta n)\pi a^2 dz = -\frac{\hat{e} M_0}{\hat{c}} \left[1 - \frac{1}{\cosh(kL)} \right]. \tag{6.4.23}$$

When n_0 increases, k increases, $\cosh(kL)$ increases and $|Q^e|$ increases as expected. We introduce the following charge-torque coefficient

$$\gamma = \frac{Q^e}{M_0} = -\frac{\hat{e}}{\hat{c}} \left[1 - \frac{1}{\cosh(kL)} \right], \tag{6.4.24}$$

as a measure of the strength of the coupling between torsion and carrier redistribution.

6.5 Higher-Order One-Dimensional Equations

The one-dimensional equations for extension, torsion and bending of thin rods are in terms of the zero- and first-order moments over a cross-section. These equations can be obtained together systematically from the three-dimensional equations using double power series expansions over a cross-section as initiated by Mindlin for an elastic rod [41]. The power series expansion procedure for developing one-dimensional theories was extended subsequently to piezoelectric dielectric [42–44] and semiconductor [45,46] rods. In this section, we use the procedure to establish a hierarchy of one-dimensional equations for the extension, torsion and bending of a piezoelectric semiconductor rod [46]. Consider the thin rod in Fig. 6.8. x_1 and x_2 are the centroidal and principal axes within the cross-section. The equations to be derived are valid for an arbitrary cross-section A. A rectangular rod is shown in the figure. The boundary curve of the cross-section is denoted by C, whose unit outward normal is $\mathbf{n}$.

We begin with the following double power series expansions in x_1 and x_2:

$$u_i(x_1, x_2, x_3, t) = \sum_{n,m=0}^{\infty} x_1^n x_2^m u_i^{(n,m)}(x_3, t),$$

$$\varphi(x_1, x_2, x_3, t) = \sum_{n,m=0}^{\infty} x_1^n x_2^m \varphi^{(n,m)}(x_3, t),$$

$$(6.5.1)$$

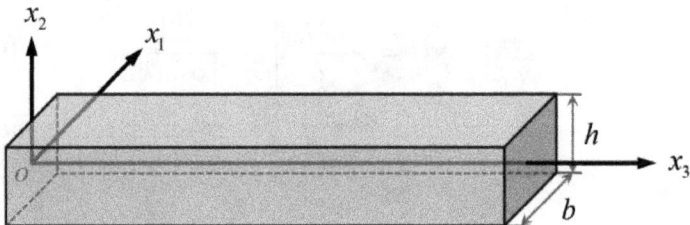

Fig. 6.8.   A thin rod and coordinate system.

$$\Delta p(x_1, x_2, x_3, t) = \sum_{n,m=0}^{\infty} x_1^n x_2^m p^{(n,m)}(x_3, t),$$

$$\Delta n(x_1, x_2, x_3, t) = \sum_{n,m=0}^{\infty} x_1^n x_2^m n^{(n,m)}(x_3, t).$$

(6.5.2)

The substitution of Eqs. (6.5.1) and (6.5.2) into the three-dimensional strain–displacement and electric field–potential relations yields the following series expansions for strains and electric fields:

$$S_{ij} = \sum_{n=0}^{\infty} x_1^n x_2^m S_{ij}^{(n,m)}, \quad E_i = \sum_{n=0}^{\infty} x_1^n x_2^m E_i^{(n,m)},$$

(6.5.3)

where

$$S_{ij}^{(n,m)} = \frac{1}{2} [\delta_{3j} u_{i,3}^{(m,n)} + \delta_{3i} u_{j,3}^{(m,n)}$$

$$+ (n+1)(\delta_{1j} u_i^{(n+1,m)} + \delta_{1i} u_j^{(n+1,m)})$$

$$+ (m+1)(\delta_{2j} u_i^{(n,m+1)} + \delta_{2i} u_j^{(n,m+1)})],$$

$$E_i^{(n,m)} = -[\delta_{3i} \varphi_{,3}^{(n,m)} + (n+1)\delta_{1i} \varphi^{(n+1,m)} + (m+1)\delta_{2i} \varphi^{(n,m+1)}].$$

(6.5.4)

Similarly, we can also write the three-dimensional carrier concentration perturbation gradients as

$$(\Delta p)_{,i} = \sum_{n=0}^{\infty} x_1^n x_2^m P_i^{(n,m)}, \quad (\Delta n)_{,i} = \sum_{n=0}^{\infty} x_1^n x_2^m N_i^{(n,m)},$$

(6.5.5)

where

$$P_i^{(n,m)} = \delta_{3i} p_{,3}^{(n,m)} + (n+1)\delta_{1i} p^{(n+1,m)} + (m+1)\delta_{2i} p^{(n,m+1)},$$

$$N_i^{(n,m)} = \delta_{3i} n_{,3}^{(n,m)} + (n+1)\delta_{1i} n^{(n+1,m)} + (m+1)\delta_{2i} n^{(n,m+1)}.$$

(6.5.6)

One-dimensional field equations are obtained by integrating the products of the three-dimensional field equations with $x_1^n x_2^m$ over the cross-section A. With the use of the two-dimensional integration by parts (divergence theorem) over A, the following equations can be obtained:

$$T_{i3,3}^{(n,m)} - nT_{i1}^{(n-1,m)} - mT_{i2}^{(n,m-1)} + F_i^{(n,m)}$$

$$= \rho \sum_{p,q=0}^{\infty} I^{(n+p,m+q)} \ddot{u}_i^{(p,q)},$$

$$D_{3,3}^{(n,m)} - nD_1^{(n-1,m)} - mD_2^{(n,m-1)} + D^{(n,m)}$$

$$= q \sum_{p,q=0}^{\infty} I^{(n+p,m+q)} (p^{(p,q)} - n^{(p,q)}),$$

(6.5.7)

$$J_{3,3}^{p(n,m)} - nJ_1^{p(n-1,m)} - mJ_2^{p(n,m-1)} + J^{p(n,m)}$$

$$= -q \sum_{p,q=0}^{\infty} I^{(n+p,m+q)} \dot{p}^{(p,q)},$$

$$J_{3,3}^{n(n,m)} - nJ_1^{n(n-1,m)} - mJ_2^{n(n,m-1)} + J^{n(n,m)}$$

$$= q \sum_{p,q=0}^{\infty} I^{(n+p,m+q)} \dot{n}^{(p,q)},$$

(6.5.8)

where the moments of inertia of A and the electromechanical resultants over the cross-section of various orders are defined by

$$\{I^{(n,m)}, T_{ij}^{(n,m)}, D_i^{(n,m)}, J_i^{p(n,m)}, J_i^{n(n,m)}\}$$

$$= \int_A \{1, T_{ij}, D_i, J_i^p, J_i^n\} x_1^n x_2^m dA.$$

(6.5.9)

In Eqs. (6.5.7) and (6.5.8), the body force and electromechanical loads on the lateral surface of the rod are represented by

$$F_i^{(n,m)} = \int_A f_i x_1^n x_2^m dA + \oint_C T_{ij} n_j x_1^n x_2^m dl,$$

$$D^{(n,m)} = \oint_C D_i n_i x_1^n x_2^m dl,$$

(6.5.10)

$$J^{p(n,m)} = \oint_C J_i^p n_i x_1^n x_2^m dl, \quad J^{n(n,m)} = \oint_C J_i^n n_i x_1^n x_2^m dl.$$

Substituting the three-dimensional linear constitutive relations in Eqs. (4.3.3) and (4.3.8) into the resultants in Eq. (6.5.9), using the series expansions in Eqs. (6.5.3) and (6.5.5), we obtain the one-dimensional constitutive relations as

$$T_{ij}^{(n,m)} = \sum_{p,q=0}^{\infty} I^{(n+p,m+q)} (c_{ijkl} S_{kl}^{(p,q)} - e_{kij} E_k^{(p,q)}),$$

$$D_i^{(n,m)} = \sum_{p,q=0}^{\infty} I^{(n+p,m+q)} (\varepsilon_{ij} E_j^{(p,q)} + e_{ijk} S^{(p,q)})_{jk},$$

(6.5.11)

$$J_i^{p(n,m)} = q \sum_{p,q=0}^{\infty} I^{(n+p,m+q)} (p_0 \mu_{ij}^p E_j^{(p,q)} - D_{ij}^p P^{(p,q)})_j,$$

$$J_i^{n(n,m)} = q \sum_{p,q=0}^{\infty} I^{(n+p,m+q)} (n_0 \mu_{ij}^n E_j^{(p,q)} D_{ij}^n N^{(p,q)})_j.$$

(6.5.12)

With successive substitutions from Eqs. (6.5.11), (6.5.12), (6.5.4) and (6.5.6), we can write Eqs. (6.5.7) and (6.5.8) as a system of one-dimensional equations for $u_i^{(n,m)}$, $\varphi^{(n,m)}$, $p^{(n,m)}$ and $n^{(n,m)}$. The zero- and first-order ones of these equations cover the usual extension, torsion and bending of thin rods. The higher-order ones can describe more sophisticated deformations of rods. Some of them will be explored in the next section.

6.6 Rectangular Piezoelectric Semiconductor Shafts

For the torsion of a rectangular shaft with consideration of warping and the shear deformation within a cross-section (in-plane shear), we begin the truncation of the series expansions in the previous section as follows [46,47]:

$$u_1 \cong x_2 u_1^{(0,1)}, \quad u_2 \cong x_1 u_2^{(1,0)}, \quad u_3 \cong x_1 x_2 u_3^{(1,1)},$$

$$\varphi \cong \varphi^{(0,0)}, \quad \Delta p \cong p^{(0,0)}, \quad \Delta n \cong n^{(0,0)},$$

(6.6.1)

where the displacement components related to in-plane shear and warping are as shown in Fig. 6.9.

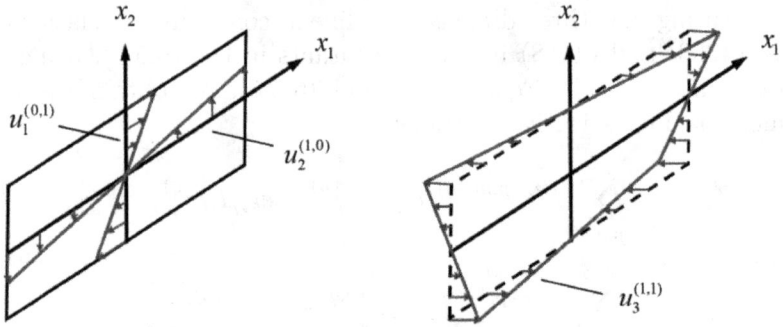

Fig. 6.9.   Displacements related to in-plane shear and warping.

Accordingly,

$$S_{33}^{(1,1)} = u_{3,3}^{(1,1)}, \quad 2S_{23}^{(1,0)} = u_3^{(1,1)} + u_{2,3}^{(1,0)},$$

$$2S_{13}^{(0,1)} = u_3^{(1,1)} + u_{1,3}^{(0,1)}, \quad 2S_{12}^{(0,0)} = u_1^{(0,1)} + u_2^{(1,0)}, \tag{6.6.2}$$

$$E_3^{(0,0)} = -\varphi_{,3}^{(0,0)}, \quad P_3^{(0,0)} = p_{,3}^{(0,0)}, \quad N_3^{(0,0)} = n_{,3}^{(0,0)}, \tag{6.6.3}$$

where $u_3^{(1,1)}$ describes warping, and $S_{12}^{(0,0)}$, the in-plane shear. The relevant one-dimensional field equations are

$$T_{13,3}^{(0,1)} - T_{12}^{(0,0)} + F_1^{(0,1)} = \rho I^{(0,2)} \ddot{u}_1^{(0,1)},$$

$$T_{23,3}^{(1,0)} - T_{12}^{(0,0)} + F_2^{(1,0)} = \rho I^{(2,0)} \ddot{u}_2^{(1,0)}, \tag{6.6.4}$$

$$T_{33,3}^{(1,1)} - T_{13}^{(0,1)} - T_{23}^{(1,0)} + F_1^{(1,1)} = \rho I^{(2,2)} \ddot{u}_3^{(1,1)},$$

$$D_{3,3}^{(0,0)} + D^{(0,0)} = qA(p^{(0,0)} - n^{(0,0)}), \tag{6.6.5}$$

$$J_{3,3}^{p(0,0)} + J^{p(0,0)} = -qA\dot{p}^{(0,0)},$$

$$J_{3,3}^{n(0,0)} + J^{n(0,0)} = qA\dot{n}^{(0,0)}, \tag{6.6.6}$$

where $I^{(0,0)} = A$ has been used. For constitutive relations, we consider cubic crystals whose material matrices are given by Eqs. (4.1.43) and (4.1.44). The relevant one-dimensional constitutive relations take the following form:

$$T_{12}^{(0,0)} = A[c_{44}(u_1^{(0,1)} + u_2^{(1,0)}) + e_{14}\varphi_{,3}^{(0,0)}],$$

$$T_{23}^{(1,0)} = I^{(2,0)}c_{44}(u_3^{(1,1)} + u_{2,3}^{(1,0)}),$$

$$T_{13}^{(0,1)} = I^{(0,2)}c_{44}(u_3^{(1,1)} + u_{1,3}^{(0,1)}),$$

$$T_{33}^{(1,1)} = I^{(2,2)}c_{11}u_{3,3}^{(1,1)},$$

(6.6.7)

$$D_3^{(0,0)} = A[-\varepsilon_{11}\varphi_{,3}^{(0,0)} + e_{14}(u_1^{(0,1)} + u_2^{(1,0)})],$$ (6.6.8)

$$J_3^{p(0,0)} = qA[-p_0\mu_{11}^p\varphi_{,3}^{(0,0)} - D_{11}^p p_{,3}^{(0,0)}],$$

$$J_3^{n(0,0)} = qA[-n_0\mu_{11}^n\varphi_{,3}^{(0,0)} + D_{11}^n n_{,3}^{(0,0)}].$$

(6.6.9)

For a physically more revealing description, we introduce the angle of twist ψ by [47] as follows:

$$\psi = \frac{1}{2}(u_2^{(1,0)} - u_1^{(0,1)}),$$

$$u_2^{(1,0)} = S_{12}^{(0,0)} + \psi, \quad u_1^{(0,1)} = S_{12}^{(0,0)} - \psi.$$

(6.6.10)

Correspondingly, we subtract and add Eq. (6.6.4)$_{1,2}$ to obtain

$$M_{,3} + m_3 = \rho(I^+\ddot{\psi} + I^-\ddot{S}_{12}^{(0,0)}),$$

$$M_{,3}^S - 2T_{12}^{(0,0)} + m^S = \rho(I^+\ddot{S}_{12}^{(0,0)} + I^-\ddot{\psi}),$$

(6.6.11)

where the torque M, the in-plane shear "moment" M^S, the corresponding mechanical loads m_3 and m^S, the cross-sectional properties I^+ and I^- are defined by

$$M = T_{23}^{(1,0)} - T_{13}^{(0,1)}, \quad M^S = T_{23}^{(1,0)} + T_{13}^{(0,1)},$$

$$m_3 = F_2^{(1,0)} - F_1^{(0,1)}, \quad m^S = F_2^{(1,0)} + F_1^{(0,1)},$$

(6.6.12)

$$I^+ = \kappa_{20}I^{(2,0)} + \kappa_{02}I^{(0,2)},$$

$$I^- = \kappa_{20}I^{(2,0)} - \kappa_{02}I^{(0,2)}, \quad \kappa_{20} \neq \kappa_{02}.$$

(6.6.13)

In Eq. (6.6.13), κ_{20} and κ_{02} are two correction factors which are necessary in the degenerate case of a square cross-section [47]. With

M^S, Eqs. (6.6.4)$_3$ take the following form:

$$T_{33,3}^{(1,1)} - M^S + F_1^{(1,1)} = \rho I^{(2,2)} \ddot{u}_3^{(1,1)}. \qquad (6.6.14)$$

Equations (6.6.7)$_{2-4}$ and (6.6.8) are then written in the following form:

$$M = c_{44}(I^- u_3^{(1,1)} + I^- S_{12,3}^{(0,0)} + I^+ \psi_{,3}),$$

$$M^S = c_{44}(I^+ u_3^{(1,1)} + I^+ S_{12,3}^{(0,0)} + I^- \psi_{,3}), \qquad (6.6.15)$$

$$T_{12}^{(0,0)} = A[2c_{44}S_{12}^{(0,0)} + e_{14}\varphi_{,3}^{(0,0)}],$$

$$D_3^{(0,0)} = A[-\varepsilon_{11}\varphi_{,3}^{(0,0)} + 2e_{14}S_{12}^{(0,0)}]. \qquad (6.6.16)$$

The substitution of Eqs. (6.6.15), (6.6.16), (6.6.7)$_1$ and (6.6.9) into Eqs. (6.6.11), (6.6.14), (6.6.5) and (6.6.6) gives six equations for the angle of twist ψ, warping $u_3^{(1,1)}$, in-plane shear $S_{12}^{(0,0)}$, $\varphi^{(0,0)}$, $p^{(0,0)}$ and $n^{(0,0)}$:

$$c_{44}(I^- u_{3,3}^{(1,1)} + I^- S_{12,33}^{(0,0)} + I^+ \psi_{,33}) + m_3 = \rho(I^+ \ddot{\psi} + I^- \ddot{S}_{12}^{(0,0)}),$$

$$c_{44}(I^+ u_{3,3}^{(1,1)} + I^+ S_{12,33}^{(0,0)} + I^- \psi_{,33})$$
$$- 2A(2c_{44}S_{12}^{(0,0)} + e_{14}\varphi_{,3}^{(0,0)}) + m^S = \rho(I^+ \ddot{S}_{12}^{(0,0)} + I^- \ddot{\psi}), \quad (6.6.17)$$

$$I^{(2,2)}c_{11}u_{3,33}^{(1,1)} - c_{44}(I^+ u_3^{(1,1)} + I^+ S_{12,3}^{(0,0)} + I^- \psi_{,3})$$
$$+ F_1^{(1,1)} = \rho I^{(2,2)} \ddot{u}_3^{(1,1)},$$

$$A(-\varepsilon_{11}\varphi_{,33}^{(0,0)} + 2e_{14}S_{12,3}^{(0,0)}) + D^{(0,0)} = qA(p^{(0,0)} - n^{(0,0)}), \quad (6.6.18)$$

$$qA[-p_0\mu_{11}^p\varphi_{,33}^{(0,0)} - D_{11}^p p_{,33}^{(0,0)}] + J^{p(0,0)} = -qA\dot{p}^{(0,0)},$$

$$qA[-n_0\mu_{11}^n\varphi_{,33}^{(0,0)} + D_{11}^n n_{,33}^{(0,0)}] + J^{n(0,0)} = qA\dot{n}^{(0,0)}. \quad (6.6.19)$$

For some basic understanding of Eqs. (6.6.17)–(6.6.19), consider standing torsional, warping and in-plane shear waves in an infinite shaft free from any electromechanical loads. The shaft is made from a p-type semiconductor with $n^{(0,0)} = 0$. Let

$$\{\psi, S_{12}^{(0,0)}\} = \{A, B\} \sin(\xi x_3) \exp(i\omega t),$$

$$\{u_3^{(1,1)}, \varphi^{(0,0)}, p^{(0,0)}\} = \{C, D, E\} \cos(\xi x_3) \exp(i\omega t), \qquad (6.6.20)$$

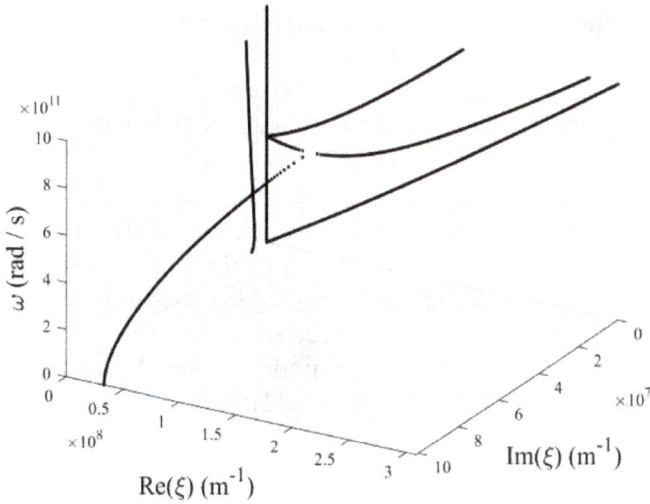

Fig. 6.10. Dispersion relations.

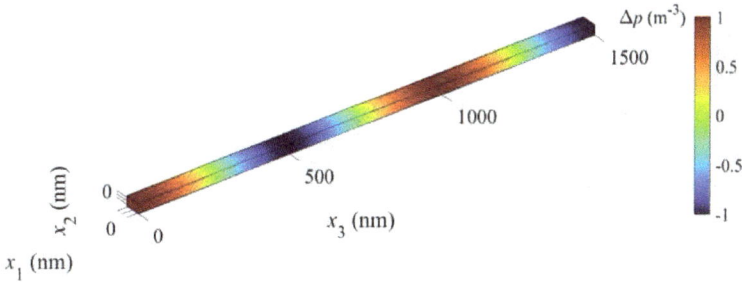

Fig. 6.11. Hole concentration perturbation distribution.

where A, B, C, D and E are undetermined constants. The substitution of Eq. (6.6.20) into Eqs. (6.6.17)–(6.6.19)$_1$ results in a system of linear homogeneous equations for A, B, C, D and E. For nontrivial solutions, the determinant of the coefficient matrix of the equations has to vanish, which leads to a relationship between ω and ξ, i.e., the dispersion relation of the waves. For numerical results, we choose a shaft of GaAs with an initial $p_0 = 10^{21}\,\mathrm{m}^{-3}$, $b = 50\,\mathrm{nm}$ and $h = 30\,\mathrm{nm}$. The dispersion relation and the hole concentration perturbation when $\xi = \pi/(500\,\mathrm{nm})$ are shown in

Figs. 6.10 and 6.11, respectively. The dispersion relation in Fig. 6.10 has multiple branches and may be complex. When ξ is approximately real, as ω increases from zero, the three branches are for torsion, in-plane shear and warping, respectively. The two upper branches for in-plane shear and warping are dispersive and have finite intercepts with the frequency axis. These intercepts are called cutoff frequencies below which the waves cannot propagate. At the minimum of the in-plane shear branch there is a complex branch. These are qualitatively similar to the dispersion curves of the elastic rectangular shaft in [47]. From Eqs. (6.6.16) it can be seen that the in-plane shear is coupled to the axial electric field under which the holes assume an axial distribution as shown in Fig. 6.11.

Chapter 7

Bending of Beams

Beams in bending or flexural deformation are widely used in devices. Qualitatively, the bending stiffness of a beam is smaller than its extensional stiffness. Therefore, bending is suitable in low-frequency applications. Theoretical modeling of bending is more complicated than that of extension. This chapter establishes one-dimensional models of various beams of functional materials in bending.

7.1 Piezoelectric Dielectric Beams with Transverse Poling

Consider the ceramic beam with transverse poling in Fig. 7.1. It is unelectroded on its lateral surfaces. The electric field in the surrounding free space is neglected. The cross-sectional area is A. The boundary curve of A is C.

For bending in the (x_1, x_3) plane with shear deformation, we make the following approximations of the relevant mechanical displacements and electric potential:

$$u_3(\mathbf{x}, t) \cong w(x_1, t),$$
$$u_1(\mathbf{x}, t) \cong x_3 \psi(x_1, t), \qquad (7.1.1)$$
$$\varphi(\mathbf{x}, t) \cong \varphi(x_1, t),$$

where $w(x_1, t)$ is the bending displacement (deflection) and $\psi(x_1, t)$, the shear displacement accompanying bending. The relevant strain

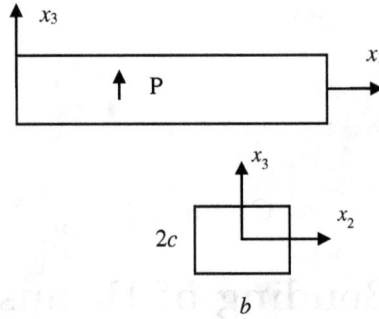

Fig. 7.1.   A ceramic beam and its rectangular cross-section.

and electric field components are

$$S_1 = u_{1,1} = x_3\psi_{,1},$$
$$S_5 = u_{3,1} + u_{1,3} = w_{,1} + \psi, \qquad (7.1.2)$$
$$E_1 = -\varphi_{,1}.$$

The main stress components are the normal stress T_1 and shear stress $T_{13} = T_5$ over a cross-section. We apply the following stress relaxation for thin beams:

$$T_2 = T_3 \cong 0. \qquad (7.1.3)$$

The constitutive relations for the relevant strain and electric displacement components are

$$S_1 = s_{11}T_1,$$
$$S_5 = s_{55}T_5 + d_{15}E_1, \qquad (7.1.4)$$
$$D_1 = d_{15}T_5 + \varepsilon_{11}E_1.$$

We invert Eq. $(7.1.4)_{1,2}$ for expressions of stresses in terms of strains and substitute the resulting expression of T_5 into Eq. $(7.1.4)_3$. This gives

$$T_1 = \bar{c}_{11}S_1 = \bar{c}_{11}x_3\psi_{,1},$$
$$T_5 = \bar{c}_{55}\kappa^2 S_5 - \bar{e}_{15}\kappa E_1 = \bar{c}_{55}\kappa^2(w_{,1} + \psi) + \bar{e}_{15}\kappa\varphi_{,1}, \qquad (7.1.5)$$
$$D_1 = \bar{e}_{15}\kappa S_5 + \bar{\varepsilon}_{11}E_1 = \bar{e}_{15}\kappa(w_{,1} + \psi) - \bar{\varepsilon}_{11}\varphi_{,1},$$

where Eq. (7.1.2) has been used. A shear correction factor κ has been introduced into Eq. (7.1.5) in the manner of Mindlin to improve the accuracy of the one-dimensional theory being developed. The way κ appears in Eq. (7.1.5) is because it is introduced in the electric enthalpy H in Eq. (4.1.3) through $S_5 \to \kappa S_5$. κ is determined by setting equal the shear vibration frequencies of the beam calculated from the one- and three-dimensional theories, which is not pursued in this book. In the following, κ will be simply set to one. The effective material constants for thin beams in Eq. (7.1.5) are defined by

$$\bar{c}_{11} = 1/s_{11}, \quad \bar{c}_{55} = 1/s_{55}, \quad \bar{e}_{15} = d_{15}/s_{55},$$
$$\bar{\varepsilon}_{11} = \varepsilon_{11} - d_{15}^2/s_{55}. \tag{7.1.6}$$

The shear force Q, bending moment M and total axial electric displacement $\hat{D}_1$ are given by the following integrals over a cross-section:

$$Q = \int_A T_5 dA = \bar{c}_{55} A(w_{,1} + \psi) + \bar{e}_{15} A\varphi_{,1}, \tag{7.1.7}$$

$$M = \int_A x_3 T_1 dA = \bar{c}_{11} I \psi_{,1}, \tag{7.1.8}$$

$$\hat{D}_1 = \int_A D_1 dA = \bar{e}_{15} A(w_{,1} + \psi) - \bar{\varepsilon}_{11} A\varphi_{,1}, \tag{7.1.9}$$

where

$$A = 2bc, \quad I = \frac{2}{3}bc^3. \tag{7.1.10}$$

From the equation of motion of the differential element of the beam in Fig. 7.2 in the x_3 direction and the moment equation about its center, the same as the derivation of Eqs. (1.8.13) and (1.8.14), we obtain

$$Q_{,1} + F_3 = \rho A \ddot{w},$$
$$M_{,1} - Q + m_2 = \rho I \ddot{\psi}, \tag{7.1.11}$$

where $F_3(x_1, t)$ is the transverse load per unit length of the beam. $m_2(x_1, t)$ is the distributed moment per unit length of the beam.

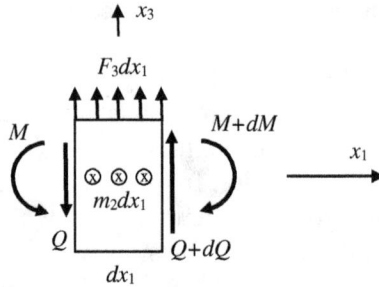

Fig. 7.2.   A differential element of the beam under mechanical loads.

For the charge equation of electrostatics, we follow a different approach this time. Instead of applying electric loads to the one-dimensional differential element in Fig. 7.2, we begin with the three-dimensional charge equation:

$$D_{k,k} = D_{1,1} + D_{2,2} + D_{3,3} = D_{1,1} + D_{b,b} = 0, \qquad (7.1.12)$$

where we have introduced a two-dimensional indicial notation and summation convention that subscript b assumes 2 and 3 but not 1. We integrate Eq. (7.1.12) over the cross-section A. This leads to

$$
\begin{aligned}
\int_A D_{1,1} dA &+ \int_A D_{b,b} dA \\
&= \left(\int_A D_1 dA \right)_{,1} + \int_C D_b n_b dl = \hat{D}_{1,1} + 0 = 0,
\end{aligned}
\qquad (7.1.13)
$$

or

$$\hat{D}_{1,1} = 0, \qquad (7.1.14)$$

where we have used the two-dimensional divergence theorem over the cross-section A and the boundary condition that $\mathbf{D} \cdot \mathbf{n} = 0$ on C, the boundary curve of A. The substitution of Eqs. (7.1.7)–(7.1.9) into Eqs. (7.1.11) and (7.1.14) yields, for a homogeneous beam, the following three equations for w, ψ and φ:

$$\bar{c}_{11}A(w_{,11}+\psi_{,1})+\bar{e}_{15}A\varphi_{,11}+F_3=\rho A\ddot{w},$$

$$\bar{c}_{11}I\psi_{,11}-[\bar{c}_{55}A(w_{,1}+\psi)+\bar{e}_{15}A\varphi_{,1}]+m_2=\rho I\ddot{\psi}, \qquad (7.1.15)$$

$$\bar{e}_{15}A(w_{,11}+\psi_{,1})-\bar{\varepsilon}_{11}A\varphi_{,11}=0.$$

At the ends of a finite beam, the following may be prescribed for boundary conditions:

$$w \quad \text{or} \quad Q, \quad \psi \quad \text{or} \quad M, \quad \varphi \quad \text{or} \quad \hat{D}_1. \qquad (7.1.16)$$

7.2 Piezoelectric Dielectric Beams with Axial Poling

Consider a piezoelectric dielectric beam of ceramics with axial poling as shown in Fig. 7.3. The lateral surface is unelectroded. The electric field in the surrounding free space is neglected as an approximation. (x,y) are the centroidal and principle axes within the cross-section A which is assumed to be symmetric about the y axis. The boundary curve of A is C. We consider bending in the (y,z) plane.

We begin with bending with shear deformation and reduce it to bending without shear deformation later. We make the following approximations of the relevant mechanical displacements and electric potential [48]:

$$u_2(\mathbf{x},t)\cong v(x_3,t), \quad u_3(\mathbf{x},t)\cong x_2\psi(x_3,t),$$
$$\varphi(\mathbf{x},t)\cong x_2\varphi^{(1)}(x_3,t), \qquad (7.2.1)$$

where $v(x_3,t)$ is the bending displacement or deflection, and $\psi(x_3,t)$, the shear displacement accompanying bending. The relevant strains

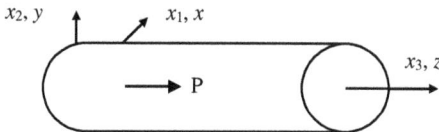

Fig. 7.3. A ceramic beam with axial poling.

and electric fields can be calculated as

$$S_3 = u_{3,3} = x_2\psi_{,3},$$
$$S_4 = u_{2,3} + u_{3,2} = v_{,3} + \psi,$$

(7.2.2)

$$E_2 = -\varphi_{,2} = -\varphi^{(1)},$$
$$E_3 = -\varphi_{,3} = -x_2\varphi^{(1)}_{,3}.$$

(7.2.3)

For bending in the (y, z) plane, the main stress components are T_3 and T_4. We introduce the following stress relaxation for thin beams:

$$T_1 = T_2 \cong 0.$$

(7.2.4)

The relevant constitutive relations are

$$S_3 = s_{33}T_3 + d_{33}E_3,$$
$$S_4 = s_{55}T_4 + d_{15}E_2,$$

(7.2.5)

$$D_2 = d_{15}T_4 + \varepsilon_{11}E_2,$$
$$D_3 = d_{33}T_3 + \varepsilon_{33}E_3.$$

(7.2.6)

For polarized ceramics, $s_{55} = s_{44}$ and $e_{15} = e_{24}$. We invert Eq. (7.2.5) for expressions of stresses in terms of strains, and substitute the resulting expressions into Eq. (7.2.6). This results in the following:

$$T_3 = \bar{c}_{33}S_3 - \bar{e}_{33}E_3 = \bar{c}_{33}x_2\psi_{,3} + \bar{e}_{33}x_2\varphi^{(1)}_{,3},$$
$$T_4 = \bar{c}_{55}S_4 - \bar{e}_{15}E_2 = \bar{c}_{55}(v_{,3} + \psi) + \bar{e}_{15}\varphi^{(1)},$$

(7.2.7)

$$D_2 = \bar{e}_{15}S_4 + \bar{\varepsilon}_{11}E_2 = \bar{e}_{15}(v_{,3} + \psi) - \bar{\varepsilon}_{11}\varphi^{(1)},$$
$$D_3 = \bar{e}_{33}S_3 + \bar{\varepsilon}_{33}E_3 = \bar{e}_{33}x_2\psi_{,3} - \bar{\varepsilon}_{33}x_2\varphi^{(1)}_{,3},$$

(7.2.8)

where Eqs. (7.2.2) and (7.2.3) have been used. The effective one-dimensional material constants in Eqs. (7.2.7) and (7.2.8) are given by

$$\bar{c}_{33} = 1/s_{33}, \quad \bar{c}_{55} = 1/s_{55},$$

$$\bar{e}_{33} = d_{33}/s_{33}, \quad \bar{e}_{15} = d_{15}/s_{55}, \tag{7.2.9}$$

$$\bar{\varepsilon}_{11} = \varepsilon_{11} - d_{15}^2/s_{55}, \quad \bar{\varepsilon}_{33} = \varepsilon_{33} - d_{33}^2/s_{33}.$$

The bending moment M, transverse shear force Q, zero- and first-order moments of the relevant electric displacements can be calculated as

$$M = \int_A x_2 T_3 dA = \bar{c}_{33} I \psi_{,3} + \bar{e}_{33} I \varphi_{,3}^{(1)},$$

$$Q = \int_A T_4 dA = \bar{c}_{55} A (v_{,3} + \psi) + \bar{e}_{15} A \varphi^{(1)}, \tag{7.2.10}$$

$$D_2^{(0)} = \int_A D_2 dA = \bar{e}_{15} A (v_{,3} + \psi) - \bar{\varepsilon}_{11} A \varphi^{(1)},$$

$$D_3^{(1)} = \int_A x_2 D_3 dA = \bar{e}_{33} I \psi_{,3} - \bar{\varepsilon}_{33} I \varphi_{,3}^{(1)}, \tag{7.2.11}$$

where I and A are the moment of inertia and the area of the beam cross-section. The one-dimensional equations of motion are obtained by applying Newton's second law and the moment equation to the differential element of the beam in Fig. 7.4. The results are the same as those of an elastic beam:

$$Q_{,3} + F_2 = \rho A \ddot{v},$$

$$M_{,3} - Q + m_1 = \rho I \ddot{\psi}. \tag{7.2.12}$$

For the charge equation of electrostatics, we begin with its three-dimensional form of

$$D_{k,k} = D_{1,1} + D_{2,2} + D_{3,3} = D_{a,a} + D_{3,3} = 0, \tag{7.2.13}$$

where we have introduced a two-dimensional indicial notation and summation convention that subscript a assumes 1 and 2 but not 3.

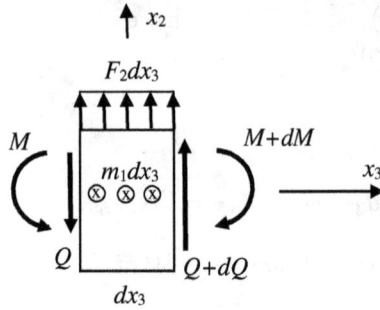

Fig. 7.4. A differential element of the beam under mechanical loads.

We multiply Eq. (7.2.13) by x_2 and integrate it over the cross-section. This leads to

$$
\int_A x_2 D_{a,a} dA + \int_A x_2 D_{3,3} dA
$$

$$
= \int_A (x_2 D_{1,1} + x_2 D_{2,2}) dA + \left(\int_A x_2 D_3 dA \right)_{,3}
$$

$$
= \int_A [(x_2 D_1)_{,1} + (x_2 D_2)_{,2} - D_2] dA + \left(\int_A x_2 D_3 dA \right)_{,3} \quad (7.2.14)
$$

$$
= \int_A [(x_2 D_a)_{,a} - D_2] dA + D_{3,3}^{(1)}
$$

$$
= \int_C x_2 D_a n_a dl - D_2^{(0)} + D_{3,3}^{(1)} = D_{3,3}^{(1)} - D_2^{(0)} = 0,
$$

or

$$
D_{3,3}^{(1)} - D_2^{(0)} = 0, \quad (7.2.15)
$$

where we have used the two-dimensional divergence theorem over the cross-section A and the boundary condition that $\mathbf{D} \cdot \mathbf{n} = 0$ on C, the boundary curve of A. The substitution of Eqs. (7.2.10) and (7.2.11) into Eqs. (7.2.12) and (7.2.15) yields, for a homogeneous beam, the following three equations for v, ψ and $\varphi^{(1)}$:

$$\bar{c}_{55}A(v_{,33} + \psi_{,3}) + \bar{e}_{15}A\varphi_{,3}^{(1)} + F_2 = \rho A\ddot{v},$$

$$\bar{c}_{33}I\psi_{,33} + \bar{e}_{33}I\varphi_{,33}^{(1)} - \bar{c}_{55}A(v_{,3} + \psi) - \bar{e}_{15}A\varphi^{(1)} + m_1$$
$$= \rho I\ddot{\psi}, \qquad\qquad\qquad (7.2.16)$$

$$\bar{e}_{33}I\psi_{,33} - \bar{\varepsilon}_{33}I\varphi_{,33}^{(1)} - \bar{e}_{15}A(v_{,3} + \psi) + \bar{\varepsilon}_{11}A\varphi^{(1)} = 0.$$

For Eq. (7.2.16), at the ends of a finite beam we may prescribe the following for boundary conditions:

$$v \quad \text{or} \quad Q, \quad \psi \quad \text{or} \quad M, \quad \varphi^{(1)} \quad \text{or} \quad D_3^{(1)}. \qquad (7.2.17)$$

The above equations are for bending with shear deformation. For the special case of bending without shear deformation, we set the beam shear strain $v_{,3} + \psi$ to zero, which implies that

$$\psi = -v_{,3}. \qquad (7.2.18)$$

Then Eqs. (7.2.10)$_1$ and (7.2.11) reduce to

$$M = -\bar{c}_{33}Iv_{,33} + \bar{e}_{33}I\varphi_{,3}^{(1)}, \qquad (7.2.19)$$

$$D_2^{(0)} = -\bar{\varepsilon}_{11}A\varphi^{(1)},$$
$$D_3^{(1)} = -\bar{e}_{33}Iv_{,33} - \bar{\varepsilon}_{33}I\varphi_{,3}^{(1)}. \qquad (7.2.20)$$

In bending without shear deformation, the moment of inertia I on the right-hand side of Eq. (7.2.12)$_2$ can be neglected. Then Eq. (7.2.12)$_2$ reduces to the following shear force-bending moment relation:

$$Q = M_{,3} + m_1 = -\bar{c}_{33}Iv_{,333} + \bar{e}_{33}I\varphi_{,33}^{(1)} + m_1, \qquad (7.2.21)$$

where Eq. (7.2.19) has been used. Effectively, Eq. (7.2.21) serves as the constitutive relation for the shear force Q, and Eq. (7.2.10)$_2$ is abandoned. When Eq. (7.2.21) is substituted into Eq. (7.2.12)$_1$, we obtain the equation for bending without shear deformation as

$$-\bar{c}_{33}Iv_{,3333} + \bar{e}_{33}I\varphi^{(1)}_{,333} + m_{1,3} + F_2 = \rho A\ddot{v}. \qquad (7.2.22)$$

Equation (7.2.22) is accompanied by Eq. (7.2.15), which takes the following form with the use of Eq. (7.2.20):

$$-\bar{e}_{33}Iv_{,333} - \bar{e}_{33}I\varphi^{(1)}_{,33} + \bar{\varepsilon}_{11}A\varphi^{(1)} = 0. \qquad (7.2.23)$$

At the ends of a finite beam, the following may be prescribed for boundary conditions:

$$v \quad \text{or} \quad Q, \quad v_{,3} \quad \text{or} \quad M, \quad \varphi^{(1)} \quad \text{or} \quad D_3^{(1)}. \qquad (7.2.24)$$

7.3   Piezoelectric Dielectric Bimorphs with Transverse Poling

Since a uniform electric field produces strains in a homogenous piezoelectric beam but not curvature, a properly constructed two-layer beam (bimorph) is often used to generate bending. Consider the ceramic bimorph in Fig. 7.5 [49,34]. When the poling direction is reversed, the elastic and dielectric constants remain the same, but the piezoelectric constants change their signs. Under a voltage across the electrodes at the top and bottom of the bimorph, one layer elongates and the other contracts along x_1 or vice versa. Hence, bending in the (x_1, x_3) plane is produced.

We want to derive one-dimensional equations for the elementary or classical bending of the bimorph without shear deformation

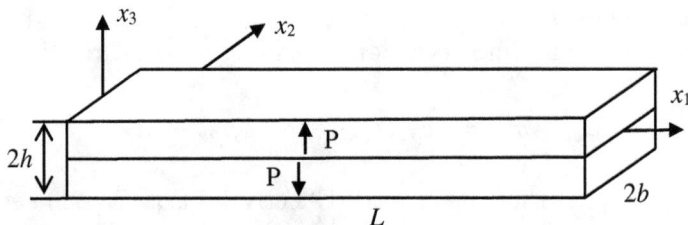

Fig. 7.5.   A ceramic bimorph with transverse poling.

(Euler–Bernoulli theory). The relevant mechanical displacements and electric potential are approximated by

$$u_3 \cong u_3(x_1, t),$$

$$u_1 \cong -x_3 u_{3,1}, \tag{7.3.1}$$

$$\varphi \cong \varphi^{(0)}(x_1, t) + x_3 \varphi^{(1)}(x_1, t).$$

On an ideal electrode with negligible resistance, the electric potential is no more than a function of time. When the top and bottom surfaces are electroded, we can write

$$\varphi = f_1(t), \quad x_3 = h,$$

$$\varphi = f_2(t), \quad x_3 = -h. \tag{7.3.2}$$

Then Eq. $(7.3.1)_3$ implies that

$$\varphi^{(0)} + h\varphi^{(1)} = f_1(t),$$

$$\varphi^{(0)} - h\varphi^{(1)} = f_2(t). \tag{7.3.3}$$

From Eq. (7.3.3), we determine

$$\varphi^{(0)} = (f_1 + f_2)/2,$$

$$\varphi^{(1)} = (f_1 - f_2)/(2h) = V(t)/(2h), \tag{7.3.4}$$

where

$$V = f_1 - f_2 \tag{7.3.5}$$

is the voltage. The electric field is given by

$$E_3 = -\varphi^{(1)} = -\frac{V}{2h}, \quad E_1 = E_2 = 0. \tag{7.3.6}$$

$\varphi^{(0)}$ does not produce any electric field. Equation $(7.3.1)_2$ implies that

$$S_1 = u_{1,1} = -x_3 u_{3,11}. \tag{7.3.7}$$

For thin beams, we make the following stress relaxation:

$$T_2 = T_3 = 0. \tag{7.3.8}$$

Then, for the upper layer,

$$S_1 = s_{11}T_1 + d_{31}E_3,$$
$$D_3 = d_{31}T_1 + \varepsilon_{33}E_3,$$
(7.3.9)

which can be rewritten as

$$T_1 = \bar{c}_{11}S_1 - \bar{e}_{31}E_3,$$
$$D_3 = \bar{e}_{31}S_1 + \bar{\varepsilon}_{33}E_3,$$
(7.3.10)

where

$$\bar{c}_{11} = \frac{1}{s_{11}}, \quad \bar{e}_{31} = \frac{d_{31}}{s_{11}}, \quad \bar{\varepsilon}_{33} = \varepsilon_{33} - \frac{d_{31}^2}{s_{11}}.$$
(7.3.11)

The one-dimensional constitutive relations for the lower layer is similar to Eq. (7.3.10), but the piezoelectric constant changes its sign.

The bending moment is calculated from

$$
\begin{aligned}
M &= \int_A T_1 x_3 \, dA = \int_A (\bar{c}_{11}S_1 \mp \bar{e}_{31}E_3)x_3 \, dA \\
&= \int_A \left(-\bar{c}_{11}x_3 u_{3,11} \pm \bar{e}_{31}\varphi^{(1)} \right) x_3 \, dA \\
&= 2b2 \int_0^h \left(-\bar{c}_{11}x_3^2 u_{3,11} + \bar{e}_{31}\varphi^{(1)} x_3 \right) dx_3 \\
&= -I\bar{c}_{11}u_{3,11} + 2bh^2 \bar{e}_{31}\varphi^{(1)},
\end{aligned}
$$
(7.3.12)

where the upper sign is for the upper layer. $A = 4bh$ is the cross-sectional area and

$$I = \frac{4bh^3}{3}.$$
(7.3.13)

The equation for the flexural motion is obtained by applying Newton's second law to the differential element in Fig. 7.6 in the x_3 direction. We have

$$(Q + dQ) - Q + F_3 dx_1 = \rho A(dx_1)\ddot{u}_3,$$
(7.3.14)

where $F_3(x_1, t)$ is the transverse mechanical load in the x_3 direction per unit length of the beam, and Q, the transverse shear force. Equation (7.3.14) can be written as

$$\frac{\partial Q}{\partial x_1} + F_3 = \rho A \ddot{u}_3.$$
(7.3.15)

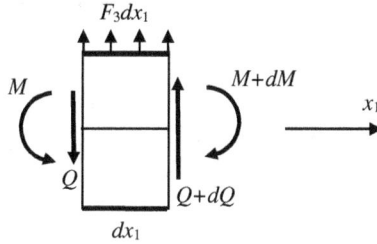

Fig. 7.6. A differential element of the bimorph.

In addition, taking moment about the center of the element in Fig. 7.6 gives

$$M + Q\frac{dx_1}{2} + (Q + dQ)\frac{dx_1}{2} - (M + dM) \cong 0, \qquad (7.3.16)$$

where the rotatory inertia of the element has been neglected. Equation (7.3.16) can be written as

$$Q = \frac{\partial M}{\partial x_1} = -I\bar{c}_{11}u_{3,111}, \qquad (7.3.17)$$

where Eq. (7.3.12) has been used. The substitution of Eq. (7.3.17) into Eq. (7.3.15) yields

$$\frac{\partial^2 M}{\partial x_1^2} + F_3 = \rho A\ddot{u}_3, \qquad (7.3.18)$$

or

$$-I\bar{c}_{11}u_{3,1111} + F_3 = \rho A\ddot{u}_3. \qquad (7.3.19)$$

Since $\varphi^{(1)}$ is no more than a function of time as shown in Eq. (7.3.4), the charge equation of electrostatics is not needed. The free charge on the top electrode at $x_3 = h$ is given by

$$Q^e = 2b \int_0^L (-D_3)|_{x_3=h}\, dx_1. \qquad (7.3.20)$$

The current flowing out of the top electrode is

$$I = -\dot{Q}^e. \qquad (7.3.21)$$

Then a circuit equation can be introduced (see Sec. 5.3).

7.4 Piezoelectric Dielectric Bimorphs with Axial Poling

Consider the case when the two layers of a bimorph have opposite axial poling directions (see Fig. 7.7) [49,34].

For the bending of the bimorph in the (x_3, x_2) plane without shear deformation, the relevant mechanical displacements and electric potential are approximated by

$$u_2 \cong u_2(x_3, t),$$
$$u_3 \cong -x_2 u_{2,3}, \tag{7.4.1}$$
$$\varphi \cong \varphi(x_3, t).$$

The axial strain and electric field are given by

$$S_3 = u_{3,3} = -x_2 u_{2,33}, \tag{7.4.2}$$
$$E_3 = -\varphi_{,3}. \tag{7.4.3}$$

For thin beams we make the following stress relaxation:

$$T_1 = T_2 = 0. \tag{7.4.4}$$

Then, for the upper layer,

$$S_3 = s_{33} T_3 + d_{33} E_3,$$
$$D_3 = d_{33} T_3 + \varepsilon_{33} E_3, \tag{7.4.5}$$

which can be inverted to give

$$T_3 = \bar{c}_{33} S_3 - \bar{e}_{33} E_3,$$
$$D_3 = \bar{e}_{33} S_3 + \bar{\varepsilon}_{33} E_3, \tag{7.4.6}$$

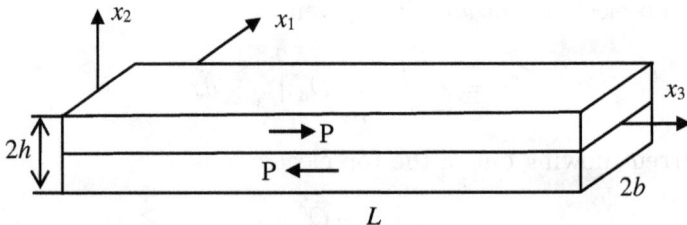

Fig. 7.7. A ceramic bimorph with axial poling.

where

$$\bar{c}_{33} = 1/s_{33}, \quad \bar{e}_{33} = d_{33}/s_{33}, \quad \bar{\varepsilon}_{33} = \varepsilon_{33} - d_{33}^2/s_{33}. \quad (7.4.7)$$

The one-dimensional constitutive relations for the lower layer are similar to Eq. (7.4.6), but the piezoelectric constant changes its sign. The bending moment is calculated from

$$
\begin{aligned}
M &= \int_A T_3 x_2 \, dA = \int_A (\bar{c}_{33} S_3 \mp \bar{e}_{33} E_3) x_2 \, dA \\
&= \int_A (-\bar{c}_{33} x_2 u_{2,33} \pm \bar{e}_{33} \varphi_{,3}) \, x_2 \, dA \\
&= 2b2 \int_0^h (-\bar{c}_{33} x_2^2 u_{2,33} + \bar{e}_{33} \varphi_{,3} x_2) \, dx_2 \\
&= -I\bar{c}_{33} u_{2,33} + 2bh^2 \bar{e}_{33} \varphi_{,3},
\end{aligned}
\quad (7.4.8)
$$

where the upper sign is for the upper layer. $A = 4bh$ is the cross-sectional area and

$$I = \frac{4bh^3}{3}. \quad (7.4.9)$$

The total axial electric displacement over a cross-section is given by

$$
\begin{aligned}
\hat{D}_3 &= \int_A D_3 dA = \int_A (\pm \bar{e}_{33} S_3 + \bar{\varepsilon}_{33} E_3) dA \\
&= \int_A (\mp \bar{e}_{33} x_2 u_{2,33} - \bar{\varepsilon}_{33} \varphi_{,3}) \, dA \\
&= 2b2 \int_0^h (-\bar{e}_{33} x_2 u_{2,33} - \bar{\varepsilon}_{33} \varphi_{,3}) \, dx_2 \\
&= -2bh^2 \bar{e}_{33} u_{2,33} - A \bar{\varepsilon}_{33} \varphi_{,3}.
\end{aligned}
\quad (7.4.10)
$$

The equation for the flexural motion is obtained by applying Newton's second law to the differential element in Fig. 7.8 in the x_2 direction. We have

$$(Q + dQ) - Q + F_2 dx_3 = \rho A (dx_3) \ddot{u}_2, \quad (7.4.11)$$

where $F_2(x_3, t)$ is the transverse mechanical load in the x_2 direction per unit length of the beam, and Q, the transverse shear force.

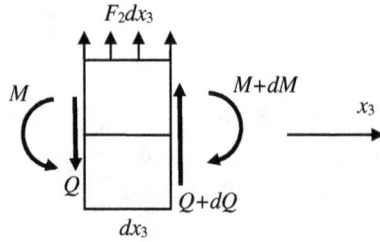

Fig. 7.8.   A differential element of the bimorph.

Equation (7.4.11) can be written as

$$\frac{\partial Q}{\partial x_3} + F_2 = \rho A \ddot{u}_2. \tag{7.4.12}$$

In addition, taking moment about the center of the element in Fig. 7.8 gives

$$M + Q\frac{dx_3}{2} + (Q + dQ)\frac{dx_3}{2} - (M + dM) \cong 0, \tag{7.4.13}$$

where the rotatory inertia of the element has been neglected. Equation (7.4.13) can be written as

$$Q = \frac{\partial M}{\partial x_3} = -I\bar{c}_{33}u_{2,333} + 2bh^2\bar{e}_{33}\varphi_{,33}, \tag{7.4.14}$$

where Eq. (7.4.8) has been used. The substitution of Eq. (7.4.14) into Eq. (7.4.12) yields

$$\frac{\partial^2 M}{\partial x_1^2} + F_3 = \rho A \ddot{u}_3, \tag{7.4.15}$$

or

$$-I\bar{c}_{33}u_{2,3333} + 2bh^2\bar{e}_{33}\varphi_{,333} + F_2 = \rho A \ddot{u}_2. \tag{7.4.16}$$

Similarly, by considering the differential element in Fig. 7.8 under electrical loads, the charge equation of electrostatic can be written as

$$\hat{D}_{3,3} = 0. \tag{7.4.17}$$

Substituting from Eq. (7.4.10), we can write Eq. (7.4.17) as

$$-2bh^2\bar{e}_{33}u_{2,333} - A\bar{\varepsilon}_{33}\varphi_{,33} = 0. \tag{7.4.18}$$

Equations (7.4.16) and (7.4.18) are the two equations needed for determining u_2 and φ.

Suppose that the two ends of the beam are electroded. The free charge on the electrode at $x_3 = L$ is given by

$$Q^e = -\hat{D}_3(L). \tag{7.4.19}$$

The current flowing out of the electrode is

$$I = -\dot{Q}^e. \tag{7.4.20}$$

Then a circuit equation can be introduced (see Sec. 5.3).

7.5 Piezoelectric Composite Beams with Transverse Poling

Composite beams with piezoelectric and elastic layers are common in smart structures. In this section, we consider the bending of a composite beam consisting of an elastic layer between two piezoelectric layers of ceramics poled along $\pm x_3$, respectively (see Fig. 7.9) [50,34]. We assume $L \gg h$ and $L \gg b \gg h'$. The reversal of the poling direction in the lower piezoelectric layer causes a sign change of its piezoelectric constants in the coordinate system shown. Under an applied voltage, one of the two ceramic layers contracts along x_1 while the other extends or vice versa, and thus bending of the composite beam in the (x_1, x_3) plane is created.

The flexural displacement is approximated by

$$u_3 \cong u_3(x_1, t). \tag{7.5.1}$$

The axial displacement for bending without shear deformation is, accordingly,

$$u_1 \cong -x_3 u_{3,1}. \tag{7.5.2}$$

Then the axial strain is given by

$$S_1 = u_{1,1} \cong -x_3 u_{3,11}. \tag{7.5.3}$$

The electric potential at the bottom electrodes of the two ceramics layers is zero. At the two top electrodes the potential is V.

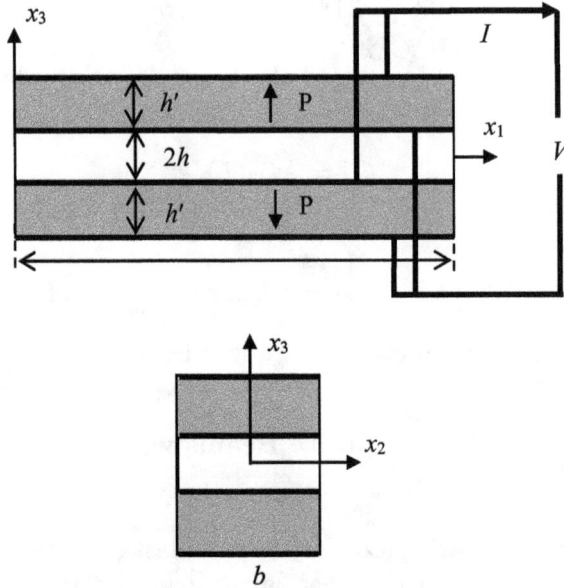

Fig. 7.9.   A composite beam with transverse poling and its cross-section.

The electric fields in the ceramic layers are given by

$$E_1 = 0, \quad E_2 = 0, \quad E_3 = -\frac{V}{h'}. \tag{7.5.4}$$

The stress–strain relation for the elastic layer is

$$T_1 = ES_1, \tag{7.5.5}$$

where E is Young's modulus. The relevant constitutive relations for the ceramic layers can be written as

$$S_1 = s_{11}T_1 \pm d_{31}E_3,$$
$$D_3 = \pm d_{31}T_1 + \varepsilon_{33}E_3, \tag{7.5.6}$$

where the stress relaxation of $T_2 = T_3 = 0$ has been used. The upper signs are for the upper layer and the lower signs for the lower layer. We solve Eq. (7.5.6) for the axial stress T_1 and the transverse electric displacement D_3 to obtain

$$T_1 = \bar{c}_{11}S_1 \mp \bar{e}_{31}E_3,$$
$$D_3 = \pm\bar{e}_{31}S_1 + \bar{\varepsilon}_{33}E_3,$$

(7.5.7)

where

$$\bar{c}_{11} = 1/s_{11}, \quad \bar{e}_{31} = d_{31}/s_{11},$$
$$\bar{\varepsilon}_{33} = \varepsilon_{33}(1 - k_{31}^2), \quad k_{31}^2 = d_{31}^2/(\varepsilon_{33}s_{11}).$$

(7.5.8)

The electric charge on the top electrode of the upper ceramic layer at $x_3 = h + h'$ is given by

$$Q^e = -b\int_0^L D_3(x_3 = h + h')dx_1.$$

(7.5.9)

Then the current in Fig. 7.9 is given by

$$I = -2\dot{Q}^e.$$

(7.5.10)

When the motion is time-harmonic, under the complex notation, we write $I = \mathrm{Re}\{\bar{I}\exp(i\omega t)\}$ and $V = \mathrm{Re}\{\bar{V}\exp(i\omega t)\}$. The electrodes may be connected to a circuit whose impedance is Z. Then, we have the following circuit equation:

$$\bar{I} = \bar{V}/Z.$$

(7.5.11)

The bending moment M is defined by the following integral over the entire cross-section of the beam:

$$M = \int_A x_3 T_1 dA = -\bar{D}u_{3,11} + \bar{e}_{31}\frac{V'}{h}2G,$$

(7.5.12)

where Eqs. (7.5.5), (7.5.7)$_1$, (7.5.2) and (7.5.4) have been used, and we have denoted

$$\bar{D} = \left\{\frac{2}{3}Eh^3 + \frac{2}{3}\bar{c}_{11}[(h + h')^3 - h^3]\right\}b,$$
$$G = \left(h + \frac{h'}{2}\right)h'b.$$

(7.5.13)

$\bar{D}$ is the bending stiffness of the beam and G is the first moment of the cross-sectional area of one of the ceramic layers about the x_2 axis.

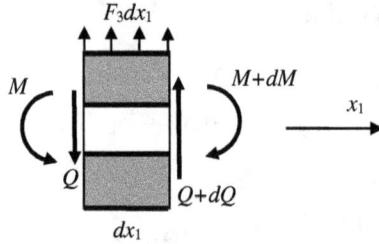

Fig. 7.10. A differential element of the beam.

The equation for flexural motion is obtained by applying Newton's second law to the differential element in Fig. 7.10 in the x_3 direction. We have

$$(Q + dQ) - Q + F_3 dx_3 = (\rho 2hb + \rho' 2h'b)(dx_1)\ddot{u}_3, \qquad (7.5.14)$$

or

$$\frac{\partial Q}{\partial x_1} + F_3 = 2b(\rho h + \rho' h')\ddot{u}_3, \qquad (7.5.15)$$

where Q is the transverse shear force, $F_3(x_1, t)$, the mechanical load per unit length of the beam in the x_3 direction, ρ and ρ', the mass densities of the elastic and piezoelectric layers.

In addition, taking moment about the center of the element in Fig. 7.10 gives

$$M + Q\frac{dx_1}{2} + (Q + dQ)\frac{dx_1}{2} - (M + dM) \cong 0, \qquad (7.5.16)$$

where the rotatory inertia of the element has been neglected. Equation (7.5.16) can be written as

$$Q = \frac{\partial M}{\partial x_1} = -\overline{D}u_{3,111}, \qquad (7.5.17)$$

where Eq. (7.5.12) has been used. The substitution of Eq. (7.5.17) into Eq. (7.5.15) yields

$$\frac{\partial M}{\partial x_1^2} + F_3 = 2b(\rho h + \rho' h')\ddot{u}_3, \qquad (7.5.18)$$

or

$$-\overline{D}u_{3,1111} + F_3 = 2b(\rho h + \rho' h')\ddot{u}_3. \qquad (7.5.19)$$

As a simple example of the actuation of the beam by a known voltage V, consider a static and free beam with $F_3 = 0$ and $M = 0$.

From Eq. (7.5.12), we obtain the beam bending curvature as

$$u_{3,11} = \left(\bar{e}_{31}\frac{V}{h'}2G\right)\Big/\bar{D}. \qquad (7.5.20)$$

7.6 Piezoelectric Composite Beams with Axial Poling

Consider the composite beam shown in Fig. 7.11 [51,34]. It consists of a pair of piezoelectric ceramic layers with axial poling and an elastic layer in the middle. The ceramic layers are identical except that their poling directions are opposite to each other. The length is much larger than the thickness $2(c+h)$ and the width b. The ceramic layers are unelectroded on their lateral surfaces. The electric field in the surrounding free space is neglected as an approximation.

Consider bending in the (x_3, x_2) plane. The flexural and axial displacements are approximately described by

$$u_2 \cong u_2(x_3, t), \quad u_3 = -x_2 u_{2,3}. \qquad (7.6.1)$$

The axial strain is expressed in terms of u_2 as

$$S_3 = u_{3,3} = -x_2 u_{2,33}. \qquad (7.6.2)$$

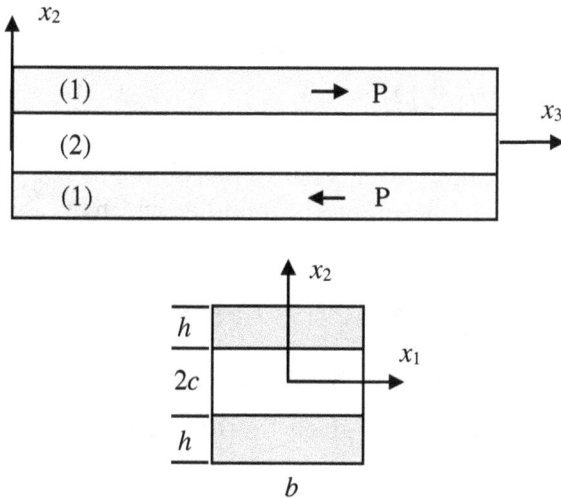

Fig. 7.11.   A composite beam with axial poling and its cross-section.

The electric potential and field are approximated by

$$\varphi \cong \varphi(x_3, t),$$
$$E_1 = E_2 = 0, \quad E_3 = -\varphi_{,3}. \tag{7.6.3}$$

Under the usual approximation of a thin beam, we have (stress relaxation)

$$T_1 = T_2 = 0. \tag{7.6.4}$$

The above approximations are valid throughout the beam.

The relevant constitutive relations for the upper ceramic layer can be written as

$$S_3 = s_{33}^{(1)} T_3 + d_{33}^{(1)} E_3,$$
$$D_3 = d_{33}^{(1)} T_3 + \varepsilon_{33}^{(1)} E_3. \tag{7.6.5}$$

We solve for the stress and electric displacement to obtain

$$T_3 = \bar{c}_{33}^{(1)} S_3 - \bar{e}_{33}^{(1)} E_3,$$
$$D_3 = \bar{e}_{33}^{(1)} S_3 + \bar{\varepsilon}_{33}^{(1)} E_3, \tag{7.6.6}$$

where

$$\bar{c}_{33}^{(1)} = 1/s_{33}^{(1)}, \quad \bar{e}_{33}^{(1)} = d_{33}^{(1)}/s_{33}^{(1)},$$
$$\bar{\varepsilon}_{33}^{(1)} = \varepsilon_{33}^{(1)}(1 - k_{33}^2), \quad k_{33}^2 = (d_{33}^{(1)})^2/(\varepsilon_{33}^{(1)} s_{33}^{(1)}). \tag{7.6.7}$$

For the lower ceramic layer with opposite poling, $d_{33}^{(1)}$ changes its sign. For the elastic layer in the middle, which is assumed to be nonpiezoelectric, the constitutive relations are

$$T_3 = \bar{c}_{33}^{(2)} S_3, \quad D_3 = \bar{\varepsilon}_{33}^{(2)} E_3,$$
$$\bar{c}_{33}^{(2)} = 1/s_{33}^{(2)}, \quad \bar{\varepsilon}_{33}^{(2)} = \varepsilon_{33}^{(2)}. \tag{7.6.8}$$

The bending moment M is given by the following integral over the cross-section of the beam:

$$M = \int_A x_2 T_3 \, dA = -\bar{D} u_{2,33} + 2G\bar{e}_{33}^{(1)} \varphi_{,3}, \tag{7.6.9}$$

where

$$\overline{D} = \left\{ \frac{2}{3}\bar{c}_{33}^{(2)} c^3 + \frac{2}{3}\bar{c}_{33}^{(1)} \left[(c+h)^3 - c^3 \right] \right\} b,$$

$$G = \left(c + \frac{h}{2} \right) hb.$$

(7.6.10)

$\overline{D}$ is the bending stiffness of the beam. G is the first moment of the cross-sectional area of one of the ceramic layers about the x_1 axis. The transverse shear force Q in the beam is defined by the integration of $T_{32} = T_4$ over a cross-section:

$$Q = \int_A T_4 \, dA. \tag{7.6.11}$$

The constitutive relation for Q is not calculated from integrating Eq. (7.6.11). Instead, it will be provided by the shear force-bending moment relation presented in what follows.

From the equation of motion of the differential element of the beam in Fig. 7.12 in the x_2 direction and the moment equation about its center, we have

$$Q_{,3} + F_2 = 2b \left(\rho^{(1)} h + \rho^{(2)} c \right) \ddot{u}_2,$$

$$M_{,3} - Q = 0,$$

(7.6.12)

where $F_2(x_3, t)$ is the transverse load per unit length of the beam. In Eq. (7.6.12)$_2$, the rotatory inertia has been neglected. Equation (7.6.12)$_2$ is the shear force-bending moment relation that serves as the constitutive relation for Q. From Eqs. (7.6.12)$_2$ and (7.6.9),

$$Q = -\overline{D}u_{2,333} + 2G\bar{e}_{33}^{(1)} \varphi_{,33}. \tag{7.6.13}$$

Eliminating Q from the two equations in Eq. (7.6.12), we obtain

$$M_{,33} + F_2 = 2b \left(\rho^{(1)} h + \rho^{(2)} c \right) \ddot{u}_2. \tag{7.6.14}$$

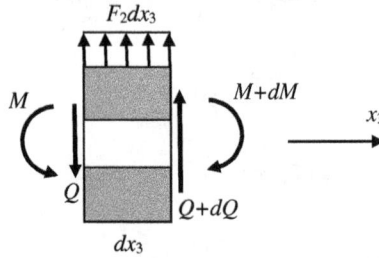

Fig. 7.12. A differential element of the beam under mechanical loads.

The total electric displacement over the cross-section of the beam is

$$\hat{D}_3 = \int_A D_3 \, dA = -2G\bar{e}_{33}^{(1)} u_{2,33} - \hat{\varepsilon}\varphi_{,3}, \qquad (7.6.15)$$

where

$$\hat{\varepsilon} = \bar{\varepsilon}_{33}^{(1)} A^{(1)} + \varepsilon_{33}^{(2)} A^{(2)},$$
$$A^{(1)} = 2bh, \quad A^{(2)} = 2bc. \qquad (7.6.16)$$

By considering the differential element in Fig. 7.12 under electric loads, the charge equation of electrostatic can be written as

$$\hat{D}_{3,3} = 0. \qquad (7.6.17)$$

When Eqs. (7.6.9) and (7.6.15) are substituted into Eqs. (7.6.14) and (7.6.17), it results in the following two equations for u_2 and φ for a homogeneous beam:

$$- \bar{D} u_{2,3333} + 2G\bar{e}_{33}^{(1)} \varphi_{,333} + F_2 = 2b \left(\rho^{(1)} h + \rho^{(2)} c \right) \ddot{u}_2,$$
$$- 2G\bar{e}_{33}^{(1)} u_{2,333} - \hat{\varepsilon}\varphi_{,33} = 0. \qquad (7.6.18)$$

7.7 Piezoelectric Dielectric Unimorphs with Transverse Poling

Unimorphs with one piezoelectric layer and one elastic layer are often used in devices operating with bending deformation. Consider the unimorph in Fig. 7.13. The piezoelectric layer is electroded at its top

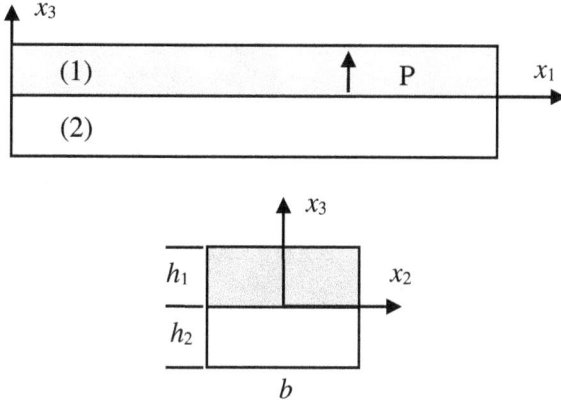

Fig. 7.13.   A unimorph with transverse poling and its cross-section.

and bottom with a voltage across the electrodes. The x_1-axis is at the interface which experiences both extension and bending in general because of the lack of structural symmetry about the interface. The x_1 axis may also be placed at the geometric middle plane of the structure, or the mass center.

For coupled extension and bending in the (x_1, x_3) plane without shear deformation, the deflection and axial displacement are approximated by

$$u_3 \cong u_3(x_1, t), \quad u_1 \cong u_1(x_1, t) - x_3 u_{3,1}. \qquad (7.7.1)$$

Then the axial strain is given by

$$S_1 = u_{1,1} - x_3 u_{3,11}. \qquad (7.7.2)$$

The electric potential and electric field in the piezoelectric layer are approximated by

$$\varphi \cong -\frac{V}{h_1} x_3, \quad E_1 = E_2 = 0, \quad E_3 = -\frac{V}{h_1}. \qquad (7.7.3)$$

For the piezoelectric layer, the relevant constitutive relations are

$$\begin{aligned} S_1 &= s_{11}^{(1)} T_1 + d_{31}^{(1)} E_3, \\ D_3 &= d_{31}^{(1)} T_1 + \varepsilon_{33}^{(1)} E_3, \end{aligned} \qquad (7.7.4)$$

where $T_2 = T_3 = 0$ (stress relaxation) has been assumed. We solve Eq. (7.7.4) for the axial stress T_1 and the transverse electric displacement D_3 to obtain

$$T_1 = \bar{c}_{11}^{(1)} S_1 - \bar{e}_{31}^{(1)} E_3 = \bar{c}_{11}^{(1)} (u_{1,1} - x_3 u_{3,11}) - \bar{e}_{31}^{(1)} E_3,$$

$$D_3 = \bar{e}_{31}^{(1)} S_1 + \bar{\varepsilon}_{33}^{(1)} E_3 = \bar{e}_{31}^{(1)} (u_{1,1} - x_3 u_{3,11}) + \bar{\varepsilon}_{33}^{(1)} E_3,$$

$$(7.7.5)$$

where

$$\bar{c}_{11}^{(1)} = 1/s_{11}^{(1)}, \quad \bar{e}_{31}^{(1)} = d_{31}^{(1)}/s_{11}^{(1)},$$

$$\bar{\varepsilon}_{33}^{(1)} = \varepsilon_{33}^{(1)} (1 - k_{31}^2), \quad k_{31}^2 = (d_{31}^{(1)})^2/(\varepsilon_{33}^{(1)} s_{11}^{(1)}).$$

$$(7.7.6)$$

For the elastic layer which is nonpiezoelectric, the relevant one-dimensional constitutive relation takes the following form:

$$T_1 = \bar{c}_{11}^{(2)} S_1 = \bar{c}_{11}^{(2)} (u_{1,1} - x_3 u_{3,11}), \qquad (7.7.7)$$

where

$$\bar{c}_{11}^{(2)} = 1/s_{11}^{(2)}. \qquad (7.7.8)$$

The total axial force N, shear force Q and bending moment M about the interface are defined by the following integrations over a cross-section:

$$N = \int_A T_1 dA = b \int_0^{h_1} \left[\bar{c}_{11}^{(1)} (u_{1,1} - x_3 u_{3,11}) - \bar{e}_{31}^{(1)} E_3 \right] dx_3$$

$$+ b \int_{-h_2}^0 \bar{c}_{11}^{(2)} (u_{1,1} - x_3 u_{3,11}) dx_3$$

$$= b(h_1 \bar{c}_{11}^{(1)} + bh_2 \bar{c}_{11}^{(2)}) u_{1,1} + b \left(b \bar{c}_{11}^{(2)} \frac{h_2^2}{2} - \bar{c}_{11}^{(1)} \frac{h_1^2}{2} \right) u_{3,11}$$

$$- bh_1 \bar{e}_{31}^{(1)} E_3,$$

$$(7.7.9)$$

$$Q = \int_A T_{13} \, dA, \qquad (7.7.10)$$

$$M = \int_A T_1 x_3 dA = b \int_0^{h_1} \left[\bar{c}_{11}^{(1)} (u_{1,1} - x_3 u_{3,11}) - \bar{e}_{31}^{(1)} E_3 \right] x_3 dx_3$$

$$+ b \int_{-h_2}^0 \bar{c}_{11}^{(2)} (u_{1,1} - x_3 u_{3,11}) x_3 dx_3$$

$$= b \left(\bar{c}_{11}^{(1)} \frac{h_1^2}{2} - \bar{c}_{11}^{(2)} \frac{h_2^2}{2} \right) u_{1,1} - b \left(\bar{c}_{11}^{(1)} \frac{h_1^3}{3} + \bar{c}_{11}^{(2)} \frac{h_2^3}{3} \right) u_{3,11}$$

$$- b \bar{e}_{31}^{(1)} \frac{h_1^2}{2} E_3. \tag{7.7.11}$$

The constitutive relation for the shear force Q will be provided by the following shear force-bending moment relation. From the equations of motion (Newton's second law) of the differential element of the unimorph in Fig. 7.14 in the x_1 and x_3 directions as well as the moment equation about the center of its interface, we have

$$N_{,1} + F_1 = b \left(\rho^{(1)} h_1 + \rho^{(2)} h_2 \right) \ddot{u}_1 + b \left(\rho^{(2)} \frac{h_2^2}{2} - \rho^{(1)} \frac{h_1^2}{2} \right) \ddot{u}_{3,1},$$

$$Q_{,1} + F_3 = b \left(\rho^{(1)} h_1 + \rho^{(2)} h_2 \right) \ddot{u}_3, \tag{7.7.12}$$

$$M_{,1} - Q + m_2 = 0,$$

where $F_1(x_1, t)$ and $F_3(x_1, t)$ are the axial and transverse loads per unit length of the beam at $x_3 = 0$. $m_2(x_1, t)$ is the distributed moment per unit length of the beam. The second term on the right-hand side of Eq. $(7.7.12)_1$ is due to the fact that the center of mass of the element in Fig. 7.14 deviates a little from the interface. When the mass densities and thicknesses of the two layers are not very different, this term is small. It is usually smaller than the first term on

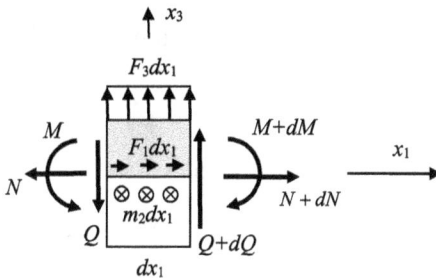

Fig. 7.14. A differential element of the unimorph under mechanical loads.

the right-hand side of Eq. $(7.7.12)_1$ because it depends on the square of the small layer thickness, and that it is involved with the displacement gradient $u_{3,1}$ which is small for long waves or fields varying slowly along x_1.

Equation $(7.7.12)_3$ is the shear force-bending moment relation. Equations $(7.7.12)_3$ and $(7.7.11)$ imply that

$$Q = M_{,1} + m_2$$
$$= b\left(\bar{c}_{11}^{(1)}\frac{h_1^2}{2} - \bar{c}_{11}^{(2)}\frac{h_2^2}{2}\right)u_{1,11} - b\left(\bar{c}_{11}^{(1)}\frac{h_1^3}{3} + \bar{c}_{11}^{(2)}\frac{h_2^3}{3}\right)u_{3,111} + m_2,$$

$$(7.7.13)$$

which serves as the constitutive relation for Q. The substitution of Eqs. $(7.7.9)$ and $(7.7.13)$ into Eq. $(7.7.12)_{1,2}$ yields two equations for u_1 and u_3.

In the special case when the unimorph is in equilibrium under a constant voltage only without mechanical loads, we have $N = 0$ and $M = 0$. Then Eqs. $(7.7.9)$ and $(7.7.11)$ reduce to

$$N = b(h_1\bar{c}_{11}^{(1)} + bh_2\bar{c}_{11}^{(2)})u_{1,1} + b\left(b\bar{c}_{11}^{(2)}\frac{h_2^2}{2} - \bar{c}_{11}^{(1)}\frac{h_1^2}{2}\right)u_{3,11}$$
$$- bh_1\bar{e}_{31}^{(1)}E_3 = 0,$$

$$(7.7.14)$$

$$M = b\left(\bar{c}_{11}^{(1)}\frac{h_1^2}{2} - \bar{c}_{11}^{(2)}\frac{h_2^2}{2}\right)u_{1,1} - b\left(\bar{c}_{11}^{(1)}\frac{h_1^3}{3} + \bar{c}_{11}^{(2)}\frac{h_2^3}{3}\right)u_{3,11}$$
$$- b\bar{e}_{31}^{(1)}\frac{h_1^2}{2}E_3 = 0.$$

$$(7.7.15)$$

Equations $(7.7.14)$ and $(7.7.15)$ determine $u_{1,1}$ and $u_{3,11}$ as constants. Then the neutral axis defined by $S_1 = 0$ can be determined from Eq. $(7.7.2)$ by setting

$$S_1 = u_{1,1} - x_3 u_{3,11} = 0. \qquad (7.7.16)$$

Hence, the location of the neutral axis is given by

$$x_3 = \frac{u_{1,1}}{u_{3,11}}. \qquad (7.7.17)$$

It the x_1 axis is placed at the neutral axis, it experiences bending only without extension. However, different from the case of a two-layer elastic beam, the neutral axis given by Eq. (7.7.17) is determined from an electric load.

The center of mass is determined by

$$\bar{x}_3 = \frac{\rho^{(1)} h_1 \frac{h_1}{2} + \rho^{(2)} h_2 \left(-\frac{h_2}{2}\right)}{\rho^{(1)} h_1 + \rho^{(2)} h_2}. \tag{7.7.18}$$

If the x_1-axis is chosen to be at the mass center, the second term on the right-hand side of Eq. $(7.7.12)_1$ disappears.

7.8   Piezoelectric Dielectric Unimorphs with Axial Poling

Consider the unimorph in Fig. 7.15 [52,34]. It consists of a piezoelectric layer (1) of ceramics poled along the axial direction and a nonpiezoelectric elastic layer (2). The unimorph is unelectroded on its lateral surfaces. The x_3-axis is at the interface which is in both extension and bending in general because of the lack of structural symmetry about the interface.

For extension and bending without shear deformation in the (x_3, x_2) plane, the bending and axial displacements are approximated

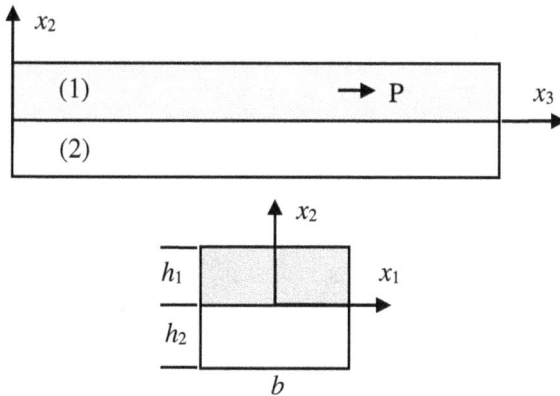

Fig. 7.15.   A unimorph with axial poling and its cross-section.

by

$$u_2 \cong v(x_3, t), \quad u_3 \cong w(x_3, t) - x_2 v_{,3}. \tag{7.8.1}$$

The axial strain is expressed in terms of w and v through

$$S_3 = w_{,3} - x_2 v_{,33}. \tag{7.8.2}$$

The electric potential and electric field are approximated by

$$\varphi \cong \varphi(x_3, t), \quad E_1 = E_2 = 0, \quad E_3 = -\varphi_{,3}. \tag{7.8.3}$$

To simplify the notation of the axial fields and material constants, we introduce

$$\begin{aligned}
S &= S_3, \quad T = T_3, \\
E &= E_3, \quad D = D_3, \\
s &= s_{33}^E, \quad d = d_{33}, \quad \varepsilon = \varepsilon_{33}^T.
\end{aligned} \tag{7.8.4}$$

For the piezoelectric layer, under the stress relaxation that $T_1 = T_2 = 0$, the relevant constitutive relations are

$$\begin{aligned}
S &= s^{(1)} T + d^{(1)} E, \\
D &= d^{(1)} T + \varepsilon^{(1)} E.
\end{aligned} \tag{7.8.5}$$

From Eq. (7.8.5), we obtain

$$\begin{aligned}
T &= \bar{c}^{(1)} S - \bar{e}^{(1)} E \\
&= \bar{c}^{(1)} (w_{,3} - x_2 v_{,33}) + \bar{e}^{(1)} \varphi_{,3}, \\
D &= \bar{e}^{(1)} S + \bar{\varepsilon}^{(1)} E \\
&= \bar{e}^{(1)} (w_{,3} - x_2 v_{,33}) - \bar{\varepsilon}^{(1)} \varphi_{,3},
\end{aligned} \tag{7.8.6}$$

where Eqs. (7.8.2) and (7.8.3) have been used, and

$$\bar{c}^{(1)} = \frac{1}{s^{(1)}}, \quad \bar{e}^{(1)} = \frac{d^{(1)}}{s^{(1)}}, \quad \bar{\varepsilon}^{(1)} = \varepsilon^{(1)} - \frac{(d^{(1)})^2}{s^{(1)}}. \tag{7.8.7}$$

Similarly, for the elastic layer which is assumed to be nonpiezoelectric, the relevant constitutive relations are

$$S = s^{(2)}T,$$
$$D = \varepsilon^{(2)}E. \tag{7.8.8}$$

Equation (7.8.8) can be rewritten into

$$T = \bar{c}^{(2)}S = \bar{c}^{(2)}(w_{,3} - x_2 v_{,33}),$$
$$D = \bar{\varepsilon}^{(2)}E = -\varepsilon^{(2)}\varphi_{,3}, \tag{7.8.9}$$

where

$$\bar{c}^{(2)} = \frac{1}{s^{(2)}}, \quad \bar{\varepsilon}^{(2)} = \varepsilon^{(2)}. \tag{7.8.10}$$

The total axial force N, shear force Q, bending moment M and electric displacement $\hat{D}_3$ are defined by the following integrations over a cross-section:

$$N = \int_A T\,dA = b\int_0^{h_1} T\,dx_2 + b\int_{-h_2}^0 T\,dx_2$$
$$= b\left(\bar{c}^{(1)}h_1 + \bar{c}^{(2)}h_2\right)w_{,3} + b\left(\frac{\bar{c}^{(2)}}{2}h_2^2 - \frac{\bar{c}^{(1)}}{2}h_1^2\right)v_{,33} \tag{7.8.11}$$
$$+ bh_1\bar{e}^{(1)}\varphi_{,3},$$

$$Q = \int_A T_{32}\,dA, \tag{7.8.12}$$

$$M = \int_A Tx_2\,dA = b\int_0^{h_1} Tx_2\,dx_2 + b\int_{-h_2}^0 Tx_2\,dx_2$$
$$= b\left(\frac{\bar{c}^{(1)}}{2}h_1^2 - \frac{\bar{c}^{(2)}}{2}h_2^2\right)w_{,3} - b\left(\frac{\bar{c}^{(1)}}{3}h_1^3 + \frac{\bar{c}^{(2)}}{3}h_2^3\right)v_{,33} \tag{7.8.13}$$
$$+ b\frac{\bar{e}^{(1)}}{2}h_1^2\varphi_{,3},$$

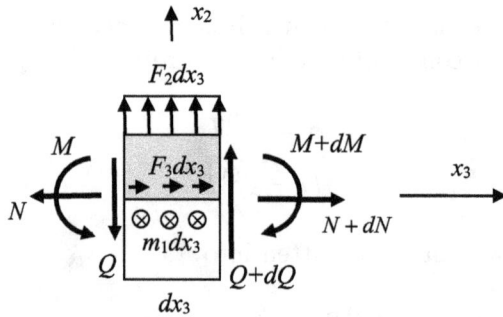

Fig. 7.16.   A differential element of the unimorph under mechanical loads.

$$\hat{D}_3 = \int_A D \, dA = b \int_0^{h_1} D \, dx_2 + b \int_{-h_2}^0 D \, dx_2$$

$$= bh_1 \bar{e}^{(1)} w_{,3} - b \frac{\bar{e}^{(1)}}{2} h_1^2 v_{,33} - b \left(h_1 \bar{\varepsilon}^{(1)} + h_2 \bar{\varepsilon}^{(2)} \right) \varphi_{,3}.$$

(7.8.14)

The constitutive relation for the shear force Q is not calculated from Eq. (7.8.12) and will be provided by the following shear force-bending moment relation.

From the equations of motion (Newton's second law) of the differential element of the unimorph in Fig. 7.16 in the x_2 and x_3 directions as well as the moment equation about the center of the interface, we have

$$N_{,3} + F_3 = b \left(\rho^{(1)} h_1 + \rho^{(2)} h_2 \right) \ddot{w} + b \left(\rho^{(2)} \frac{h_2^2}{2} - \rho^{(1)} \frac{h_1^2}{2} \right) \ddot{v}_{,3},$$

$$Q_{,3} + F_2 = b \left(\rho^{(1)} h_1 + \rho^{(2)} h_2 \right) \ddot{v},$$

(7.8.15)

$$M_{,3} - Q + m_1 = 0,$$

where $F_2(x_3, t)$ and $F_3(x_3, t)$ are the transverse and axial loads per unit length of the unimorph at $x_2 = 0$. $m_1(x_3, t)$ is the distributed moment per unit length of the unimorph. The second term on the right-hand side of Eq. (7.8.15)$_1$ is due to the fact that the center of mass of the element in Fig. 7.16 deviates a little from the interface.

Equation $(7.8.15)_3$ is the shear force-bending moment relation. Equations $(7.8.15)_3$ and $(7.8.13)$ imply that

$$
Q = b \left(\frac{\bar{c}^{(1)}}{2} h_1^2 - \frac{\bar{c}^{(2)}}{2} h_2^2 \right) w_{,33} - b \left(\frac{\bar{c}^{(1)}}{3} h_1^3 + \frac{\bar{c}^{(2)}}{3} h_2^3 \right) v_{,333}
$$
$$
+ b \frac{\bar{e}^{(1)}}{2} h_1^2 \varphi_{,33} + m_1,
$$

$(7.8.16)$

which serves as the constitutive relation for the shear force Q. Similar to the derivation of Eq. $(7.8.15)$, by considering the differential element in Fig. 7.16 under electrical loads, the charge equation of electrostatics can be written as

$$
\hat{D}_{3,3} = 0. \qquad (7.8.17)
$$

The substitution of Eqs. $(7.8.11)$, $(7.8.16)$ and $(7.8.14)$ into Eqs. $(7.8.15)_{1,2}$ and $(7.8.17)$ yields three equations for w, v and φ.

7.9 Piezoelectric Semiconductor Beams with Transverse Poling

In this section, we analyze the bending of a composite piezoelectric semiconductor beam [53,32] (see Fig. 7.17). It consists of a piezoelectric dielectric middle layer labeled by "(1)" such as ceramics poled along the x_3 direction, and two identical nonpiezoelectric semiconductor layers labeled by "(2)" such as silicon. The piezoelectric layer is in the middle where the shear stress over a cross-section is large. The piezoelectric layer is unelectroded on its lateral surfaces. The electric field in the surrounding free space is neglected.

To develop a one-dimensional model for bending in the (x_1, x_3) plane with shear deformation, we make the following approximations of the relevant mechanical displacements and electric potential:

$$
u_3(\mathbf{x}, t) \cong w(x_1, t), \quad u_1(\mathbf{x}, t) \cong x_3 \psi(x_1, t),
$$
$$
\varphi(\mathbf{x}, t) \cong \varphi(x_1, t),
$$

$(7.9.1)$

where $w(x_1, t)$ is the bending displacement (deflection) and $\psi(x_1, t)$, the shear displacement accompanying bending. The relevant strain

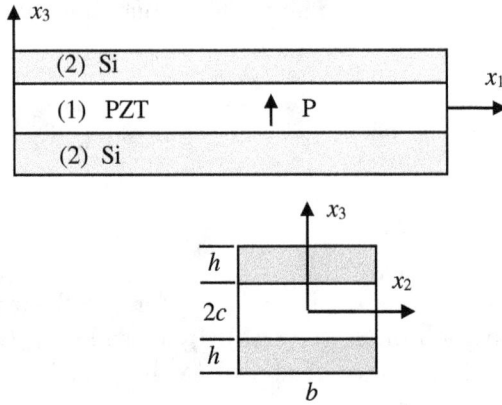

Fig. 7.17.　Side view and cross-section of a composite beam.

and electric field components are

$$S_1 = u_{1,1} = x_3 \psi_{,1},$$
$$S_5 = u_{3,1} + u_{1,3} = w_{,1} + \psi, \qquad (7.9.2)$$
$$E_1 = -\varphi_{,1}.$$

For bending in the (x_1, x_3) plane, the main stress components are the normal stress T_1 and shear stress $T_{13} = T_5$. We introduce the following stress relaxation for thin beams:

$$T_2 = T_3 \cong 0. \qquad (7.9.3)$$

For the piezoelectric layer, the constitutive relations of the relevant strain and electric displacement components are

$$S_1 = s_{11}^{(1)} T_1, \quad S_5 = s_{55}^{(1)} T_5 + d_{15} E_1,$$
$$D_1 = d_{15} T_5 + \varepsilon_{11}^{(1)} E_1. \qquad (7.9.4)$$

Equation (7.9.4) can be rewritten as

$$T_1 = T_{11} = \bar{c}_{11}^{(1)} S_1 = \bar{c}_{11}^{(1)} x_3 \psi_{,1},$$
$$T_5 = T_{31} = \bar{c}_{55}^{(1)} S_5 - \bar{e}_{15}^{(1)} E_1 = \bar{c}_{55}^{(1)} (w_{,1} + \psi) + \bar{e}_{15}^{(1)} \varphi_{,1}, \qquad (7.9.5)$$
$$D_1 = \bar{e}_{15}^{(1)} S_5 + \bar{\varepsilon}_{11}^{(1)} E_1 = \bar{e}_{15}^{(1)} (w_{,1} + \psi) - \bar{\varepsilon}_{11}^{(1)} \varphi_{,1},$$

where Eq. (7.9.2) has been used. The one-dimensional effective material constants for thin beams in Eq. (7.9.5) are defined by

$$\bar{c}_{11}^{(1)} = 1/s_{11}^{(1)}, \quad \bar{c}_{55}^{(1)} = 1/s_{55}^{(1)},$$

$$\bar{e}_{15}^{(1)} = d_{15}^{(1)}/s_{55}^{(1)}, \quad \bar{\varepsilon}_{11}^{(1)} = \varepsilon_{11}^{(1)} - (d_{15}^{(1)})^2/s_{55}^{(1)}. \tag{7.9.6}$$

Similarly, for the semiconductor layers, the one-dimensional constitutive relations for thin beams are

$$T_1 = \bar{c}_{11}^{(2)} S_1 = \bar{c}_{11}^{(2)} x_3 \psi_{,1},$$

$$T_5 = \bar{c}_{55}^{(2)} S_5 = \bar{c}_{55}^{(2)} (w_{,1} + \psi), \tag{7.9.7}$$

$$D_1 = \bar{\varepsilon}_{11}^{(2)} E_1 = -\bar{\varepsilon}_{11}^{(2)} \varphi_{,1},$$

where

$$\bar{c}_{11}^{(2)} = 1/s_{11}^{(2)}, \quad \bar{c}_{55}^{(2)} = 1/s_{55}^{(2)}, \quad \bar{\varepsilon}^{(2)} = \varepsilon_{11}^{(2)}. \tag{7.9.8}$$

The carrier concentration perturbations in the semiconductor layers are approximated by

$$\Delta p(\mathbf{x}, t) \cong \Delta p(x_1, t), \quad \Delta n(\mathbf{x}, t) \cong \Delta n(x_1, t). \tag{7.9.9}$$

The relevant one-dimensional linear constitutive relations for the current densities are

$$J_1^p = qp_0\mu_{11}^p E_1 - qD_{11}^p(\Delta p)_{,1}$$

$$= -qp_0\mu_{11}^p \varphi_{,1} - qD_{11}^p(\Delta p)_{,1},$$

$$J_1^n = qn_0\mu_{11}^n E_1 + qD_{11}^n(\Delta n)_{,1} \tag{7.9.10}$$

$$= -qn_0\mu_{11}^n \varphi_{,1} + qD_{11}^n(\Delta n)_{,1}.$$

The shear force Q, bending moment M and total axial electric displacement $\hat{D}$ are given by the following integrals over a cross-section:

$$Q = \int_A T_{13} dA = \hat{c}(w_{,1} + \psi) + \hat{e}\varphi_{,1}, \tag{7.9.11}$$

$$M = \int_A x_3 T_1 dA = \overline{D}\psi_{,1}, \tag{7.9.12}$$

$$\hat{D} = \int_A D_1 dA = \hat{e}(w_{,1} + \psi) - \hat{\varepsilon}\varphi_{,1}, \tag{7.9.13}$$

where

$$A^{(1)} = 2bc, \quad A^{(2)} = 2bh, \quad \hat{e} = \bar{e}_{15}^{(1)} A^{(1)},$$

$$\hat{c} = \bar{c}_{55}^{(1)} A^{(1)} + \bar{c}_{55}^{(2)} A^{(2)}, \quad \hat{\varepsilon} = \bar{\varepsilon}_{11}^{(1)} A^{(1)} + \bar{\varepsilon}_{11}^{(2)} A^{(2)}, \quad (7.9.14)$$

$$\overline{D} = \frac{2}{3} bc^3 \bar{\varepsilon}_{11}^{(1)} + \frac{2}{3} b[(c+h)^3 - c^3] \bar{\varepsilon}_{11}^{(2)}.$$

From the equation of motion of the differential element of the beam in Fig. 7.18 in the x_3 direction and its moment equation, we obtain

$$Q_{,1} + F_3(x_1, t) = 2b\left(\rho^{(1)} c + \rho^{(2)} h\right) \ddot{u}_3,$$

$$M_{,1} - Q = \overline{I}\ddot{\psi}, \quad (7.9.15)$$

where $F_3(x_1, t)$ is the transverse load per unit length of the beam, and

$$\overline{I} = \frac{2}{3} bc^3 \rho^{(1)} + \frac{2}{3} b[(c+h)^3 - c^3] \rho^{(2)}, \quad (7.9.16)$$

which is the rotatory inertia. Similarly, by considering the differential element in Fig. 7.18 under electric loads, the charge equation of electrostatics can be written as

$$\hat{D}_{,1} = q(\Delta p - \Delta n) A^{(2)}. \quad (7.9.17)$$

The one-dimensional conservation of holes and electrons are

$$q\frac{\partial}{\partial t}(\Delta p) = -J_{1,1}^p,$$

$$q\frac{\partial}{\partial t}(\Delta n) = J_{1,1}^n. \quad (7.9.18)$$

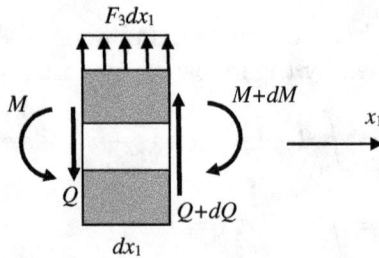

Fig. 7.18.   A differential element of the beam under mechanical loads.

The substitution of Eqs. (7.9.10)–(7.9.13) into Eqs. (7.9.15), (7.9.17) and (7.9.18) yields five equations for u_3, ψ, φ, Δp and Δn.

7.10 Piezoelectric Semiconductor Beams with Axial Poling

In this section, we consider the elementary bending of the composite beam in Fig. 7.19 [51,32] without shear deformation. The beam consists of a nonpiezoelectric semiconductor layer such as Si and a pair of piezoelectric dielectric layers such as ceramics poled along the axial direction. The two ceramic layers are identical except that their poling directions are opposite to each other. The piezoelectric layers are unelectroded on their lateral surfaces. The electric field in the surrounding free space is neglected as usual.

Consider bending without shear deformation in the (x_3, x_2) plane. The flexural and axial displacements are approximately described by

$$u_2 \cong u_2(x_3, t), \quad u_3 \cong -x_2 u_{2,3}. \tag{7.10.1}$$

Thus, the axial strain S_3 can be expressed in terms of u_2 as

$$S_3 = u_{3,3} = -x_2 u_{2,33}. \tag{7.10.2}$$

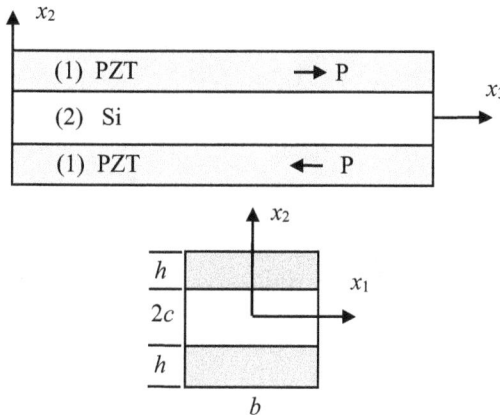

Fig. 7.19.   Side view and cross-section of a composite beam.

The electric potential and field are approximated by

$$\varphi \cong \varphi(x_3, t), \quad E_1 = E_2 = 0, \quad E_3 = -\varphi_{,3}. \qquad (7.10.3)$$

Under the usual stress approximation for a beam in bending, we assume

$$T_1 \cong 0, \quad T_2 \cong 0. \qquad (7.10.4)$$

Then the relevant constitutive relations for the upper ceramic layer can be written as

$$
\begin{aligned}
S_3 &= s_{33}^{(1)} T_3 + d_{33}^{(1)} E_3, \\
D_3 &= d_{33}^{(1)} T_3 + \varepsilon_{33}^{(1)} E_3.
\end{aligned}
\qquad (7.10.5)
$$

From Eq. (7.10.5), we solve for the stress and electric displacement to obtain

$$
\begin{aligned}
T_3 &= \bar{c}_{33}^{(1)} S_3 - \bar{e}_{33}^{(1)} E_3, \\
D_3 &= \bar{e}_{33}^{(1)} S_3 + \bar{\varepsilon}_{33}^{(1)} E_3,
\end{aligned}
\qquad (7.10.6)
$$

where

$$
\begin{aligned}
\bar{c}_{33}^{(1)} &= 1/s_{33}^{(1)}, \quad \bar{e}_{33}^{(1)} = d_{33}^{(1)}/s_{33}^{(1)}, \\
\bar{\varepsilon}_{33}^{(1)} &= \varepsilon_{33}^{(1)}(1 - k_{33}^2), \quad k_{33}^2 = (d_{33}^{(1)})^2/(\varepsilon_{33}^{(1)} s_{33}^{(1)}).
\end{aligned}
\qquad (7.10.7)
$$

Similarly, for the lower ceramic layer with opposite poling, formally Eqs. (7.10.5)–(7.10.7) are still valid, but $d_{33}^{(1)}$ changes its sign. For the semiconductor layer in the middle which is assumed to be nonpiezoelectric, the constitutive relations are

$$
\begin{aligned}
T_3 &= \bar{c}_{33}^{(2)} S_3, \quad D_3 = \bar{\varepsilon}_{33}^{(2)} E_3, \\
\bar{c}_{33}^{(2)} &= 1/s_{33}^{(2)}, \quad \bar{\varepsilon}_{33}^{(2)} = \varepsilon_{33}^{(2)}.
\end{aligned}
\qquad (7.10.8)
$$

The bending moment M is defined by the following integral over a cross-section of the beam:

$$
M = \int_A x_2 T_3 \, dA = -\bar{D} u_{2,33} + 2G \bar{e}_{33}^{(1)} \varphi_{,3}, \qquad (7.10.9)
$$

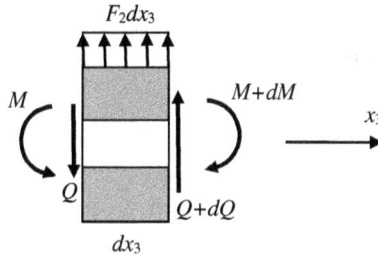

Fig. 7.20.   A differential element of the beam under mechanical loads.

where Eqs. (7.10.2) and (7.10.3) have been used, and

$$\overline{D} = \left\{ \frac{2}{3}\overline{c}_{33}^{(2)} c^3 + \frac{2}{3}\overline{c}_{33}^{(1)} \left[(c+h)^3 - c^3 \right] \right\} b,$$

$$G = \left(c + \frac{h}{2} \right) hb.$$

(7.10.10)

$\overline{D}$ is the bending stiffness of the beam. G is the first moment of the cross-sectional area of one of the ceramic layers about the x_1-axis. The transverse shear force Q in the beam is defined by the integration of T_{32} over a cross-section:

$$Q = \int_A T_{32} \, dA. \qquad (7.10.11)$$

In elementary or classical bending without shear deformation, the constitutive relation for Q is not derived from its definition in Eq. 7.10.11. Instead, it will be provided by the shear force-bending moment relation given as follows. From the equation of motion of the differential element of the beam in Fig. 7.20 in the x_2 direction and its moment equation, we obtain

$$Q_{,3} + F_2(x_3, t) = 2b \left(\rho^{(1)} h + \rho^{(2)} c \right) \ddot{u}_2,$$

$$M_{,3} - Q = 0,$$

(7.10.12)

where $F_2(x_3, t)$ is the transverse load per unit length of the beam. In Eq. (7.10.12)$_2$, the rotatory inertia of the element has been neglected.

Equation $(7.10.12)_2$ is the shear force-bending moment relation. From Eq. $(7.10.12)_2$ and Eq. $(7.10.9)$, we have

$$Q = -\overline{D}u_{2,333} + 2G\overline{e}_{33}^{(1)}\varphi_{,33}. \qquad (7.10.13)$$

The total electric displacement over a cross-section of the beam is

$$\hat{D}_3 = \int_A D_3 dA = -2G\overline{e}_{33}^{(1)}u_{2,33} - \hat{\varepsilon}\varphi_{,3}, \qquad (7.10.14)$$

where Eqs. $(7.10.6)$ and $(7.10.8)$ have been used, and

$$\hat{\varepsilon} = \overline{\varepsilon}_{33}^{(1)}A^{(1)} + \overline{\varepsilon}_{33}^{(2)}A^{(2)},$$
$$A^{(1)} = 2bh, \quad A^{(2)} = 2bc. \qquad (7.10.15)$$

Similar to the derivation of Eq. $(7.10.12)$, by considering the differential element in Fig. 7.20 under electrical loads, the charge equation of electrostatics can be written as

$$\hat{D}_{3,3} = q(\Delta p - \Delta n)A^{(2)}. \qquad (7.10.16)$$

For the one-dimensional currents, we use the linearized constitutive relations with uniform doping, as follows:

$$J_3^p \cong qp_0\mu_{33}^p E_3 - qD_{33}^p(\Delta p)_{,3},$$
$$J_3^n \cong qn_0\mu_{33}^n E_3 + qD_{33}^n(\Delta n)_{,3}. \qquad (7.10.17)$$

The one-dimensional conservation of holes and electrons are

$$q\frac{\partial}{\partial t}(\Delta p) = -J_{3,3}^p,$$
$$q\frac{\partial}{\partial t}(\Delta n) = J_{3,3}^n. \qquad (7.10.18)$$

The substitution of Eqs. $(7.10.13)$, $(7.10.14)$ and $(7.10.17)$ into Eqs. $(7.10.12)_1$, $(7.10.16)$ and $(7.10.18)$ leads to four equations for u_2, φ, Δp and Δn.

7.11 Piezomagnetic–Piezoelectric Semiconductor Beams

The composite beam under consideration in this section consists of two piezoelectric semiconductor layers such as ZnO and two piezo-magnetic (PM) layers (see Fig. 7.21) [54]. The two ZnO layers are identical except that their c-axes are opposite. The same is true for the two PM layers. The global coordinate system (x, y, z) corresponds to (x_1, x_2, x_3). The primed local coordinate system is for the upper PM layer only. The beam is under a transverse magnetic field H along the thickness direction $(3')$ of the PM layers. H is assumed to be known and uniform in the PM layers. It produces extension in the upper PM layer through the piezomagnetic constant h_{31} and contraction in the lower PM layer or vice versa, and thus generates bending of the composite beam. The ZnO layers are unelectroded on its lateral surfaces. The electric field outside the ZnO layers are neglected as an approximation.

Consider the elementary bending without shear deformation in the (y, z) plane. The flexural displacement and axial strain throughout the beam are approximately described by

$$u_2 \cong u_2(x_3, t), \quad S_3 = -x_2 u_{2,33}. \tag{7.11.1}$$

The electric potential and field as well as carrier concentration perturbations in the ZnO layers are approximated by

$$\varphi \cong \varphi(x_3, t), \quad E_1 = E_2 = 0, \quad E_3 = -\varphi_{,3},$$
$$\Delta p \cong \Delta p(x_3, t), \quad \Delta n \cong \Delta n(x_3, t). \tag{7.11.2}$$

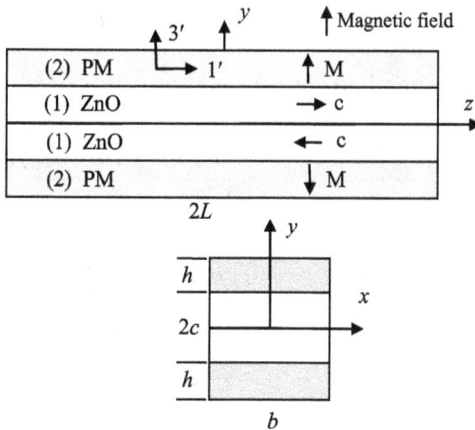

Fig. 7.21. Side view and cross-section of a composite beam.

For the upper ZnO layer, under $T_1 = T_2 = 0$, the one-dimensional constitutive relations for the axial fields are (see Eqs. (5.8.7)–(5.8.10)):

$$T = T_3 = \bar{c}^{(1)} S - \bar{e}^{(1)} E = -\bar{c}^{(1)} x_2 u_{2,33} + \bar{e}^{(1)} \varphi_{,3},$$

$$D = D_3 = \bar{e}^{(1)} S + \bar{\varepsilon}^{(1)} E = -\bar{e}^{(1)} x_2 u_{2,33} - \bar{\varepsilon}^{(1)} \varphi_{,3}, \qquad (7.11.3)$$

$$J^p = J_3^p \cong q p_0 \mu^p E - q D^p \frac{\partial(\Delta p)}{\partial z},$$

$$J^n = J_3^n \cong q n_0 \mu^n E + q D^n \frac{\partial(\Delta n)}{\partial z}, \qquad (7.11.4)$$

where Eqs. (7.11.1) and (7.11.2) have been used, and

$$\bar{c}^{(1)} = c_{33}^{(1)} - \frac{2(c_{13}^{(1)})^2}{c_{11}^{(1)} + c_{12}^{(1)}}, \quad \bar{e}^{(1)} = e_{33}^{(1)} - \frac{2 c_{13}^{(1)} e_{31}^{(1)}}{c_{11}^{(1)} + c_{12}^{(1)}},$$

$$\bar{\varepsilon}^{(1)} = \varepsilon_{33}^{(1)} + \frac{2(e_{31}^{(1)})^2}{c_{11}^{(1)} + c_{12}^{(1)}}. \qquad (7.11.5)$$

For the lower ZnO layer, the only difference is that its piezoelectric constant changes its sign. For the upper PM layer, under $T_2' = T_3' = 0$ in the local coordinate system, the effective one-dimensional constitutive relation for the axial stress is

$$T = \bar{c}^{(2)} S - \bar{h}^{(2)} H = -\bar{c}^{(2)} x_2 u_{2,33} - \bar{h}^{(2)} H, \qquad (7.11.6)$$

where Eq. (7.11.1) has been used and

$$\bar{c}^{(2)} = c_{11}^{(2)} + \frac{2 c_{12}^{(2)} (c_{13}^{(2)})^2 - c_{33}^{(2)} (c_{12}^{(2)})^2 - c_{11}^{(2)} (c_{13}^{(2)})^2}{c_{11}^{(2)} c_{33}^{(2)} - (c_{13}^{(2)})^2}, \qquad (7.11.7)$$

$$\bar{h}^{(2)} = h_{31}^{(2)} - \frac{c_{12}^{(2)} c_{33}^{(2)} h_{31}^{(2)} - c_{12}^{(2)} c_{13}^{(2)} h_{33}^{(2)} + c_{13}^{(2)} c_{11}^{(2)} h_{33}^{(2)} - (c_{13}^{(2)})^2 h_{31}^{(2)}}{c_{11}^{(2)} c_{33}^{(2)} - (c_{13}^{(2)})^2}.$$

$$\qquad (7.11.8)$$

For the lower PM layer, the piezomagnetic constant changes its sign.

For the composite beam, the bending moment M is defined by the following integral over a cross-section:

$$M = \int_A x_2 T_3 \, dA = b \int_{-(c+h)}^{c+h} x_2 T_3 \, dx_2$$

$$= 2b \int_0^c x_2 T_3 \, dx_2 + 2b \int_c^{c+h} x_2 T_3 \, dx_2 = -\overline{D} u_{2,33} + \hat{e} \varphi_{,3} - \hat{h} H,$$

$$(7.11.9)$$

where

$$\overline{D} = \left\{ \bar{c}^{(1)} \frac{2b}{3} c^3 + \bar{c}^{(2)} \frac{2b}{3} \left[(c+h)^3 - c^3 \right] \right\}, \qquad (7.11.10)$$

$$\hat{e} = \bar{e}^{(1)} bc^2, \quad \hat{h} = \bar{h}^{(2)} bh(2c + h). \qquad (7.11.11)$$

$\overline{D}$ is the bending stiffness of the composite beam. The transverse shear force Q in the beam is defined by the integration of T_{32} over a cross-section:

$$Q = \int_A T_{32} \, dA. \qquad (7.11.12)$$

The constitutive relation for Q is not derived from Eq. (7.11.12). Instead, it is provided by the shear force-bending moment relation presented in what follows. From the equation of motion of the differential element of the beam in Fig. 7.22 in the y direction and its moment equation, we obtain

$$Q_{,3} + F_2 = 2b \left(\rho^{(1)} c + \rho^{(2)} h \right) \ddot{u}_2,$$

$$(7.11.13)$$

$$M_{,3} - Q = 0,$$

where $F_2(x_3, t)$ is the transverse load per unit length of the beam.

In Eq. (7.11.13)$_2$, the rotatory inertia of the element has been neglected. Equation (7.11.13)$_2$ is the shear force-bending moment relation which produces the constitutive relation needed for Q.

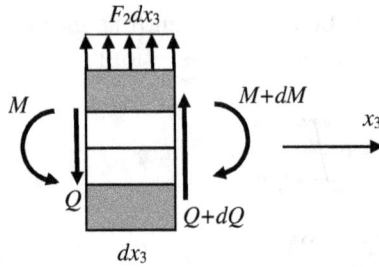

Fig. 7.22.   A differential element of the beam under mechanical loads.

From Eqs. $(7.11.13)_2$ and $(7.11.9)$, we obtain

$$Q = -\overline{D}u_{2,333} + \hat{e}\varphi_{,33}. \qquad (7.11.14)$$

The axial electric displacement over the cross-section of the two ZnO layers together is

$$\hat{D}_3 = b \int_{-c}^{c} D_3 \, dx_2 = -\hat{e}u_{2,33} - \hat{\varepsilon}\varphi_{,3}, \qquad (7.11.15)$$

where

$$\hat{\varepsilon} = \overline{\varepsilon}^{(1)}A^{(1)}, \quad A^{(1)} = 2bc. \qquad (7.11.16)$$

Similar to the derivation of Eq. $(7.11.13)$, by considering the differential element in Fig. 7.22 under electric loads, the charge equation of electrostatics can be written as

$$\hat{D}_{3,3} = q(\Delta p - \Delta n)A^{(1)}. \qquad (7.11.17)$$

The one-dimensional conservation of holes and electrons are

$$q\frac{\partial}{\partial t}(\Delta p) = -J^{p}_{3,3},$$

$$q\frac{\partial}{\partial t}(\Delta n) = J^{n}_{3,3}. \qquad (7.11.18)$$

The substitution of Eqs. $(7.11.14)$, $(7.11.15)$ and $(7.11.4)$ into Eqs. $(7.11.13)_1$, $(7.11.17)$ and $(7.11.18)$ yields four equations for u_2, φ, Δp and Δn.

As an example, consider the static bending of a beam within $|z| < L$. It is under H only without electromechanical loads. The boundary conditions are

$$M(\pm L) = 0, \quad Q(\pm L) = 0, \quad \hat{D}_3(\pm L) = 0,$$
$$J^p(\pm L) = 0, \quad J^n(\pm L) = 0. \tag{7.11.19}$$

We also have the following charge neutrality conditions:

$$\int_{-L}^{L} \Delta p \, dz = 0, \quad \int_{-L}^{L} \Delta n \, dz = 0. \tag{7.11.20}$$

Only one of Eq. (7.11.20) is independent. To make the displacement and potential fields unique, we specify

$$u_2(0) = 0, \quad u_{2,3}(0) = 0, \quad \varphi(0) = 0. \tag{7.11.21}$$

Mathematically, we have a system of linear ordinary differential equations with constant coefficients. Its solution can be obtained through a standard procedure. The results are

$$u_2 = \frac{\hat{e}^2 \hat{h}}{\overline{D}^2 \bar{\varepsilon}} \frac{H \cos h(kx_3)}{k^2 \cos h(kL)} - \frac{\hat{h}H}{2\overline{D}} x_3^2 - \frac{\hat{e}^2 \hat{h}}{\overline{D}^2 \bar{\varepsilon}} \frac{H}{k^2 \cos h(kL)}, \tag{7.11.22}$$

$$\varphi = \frac{\hat{e}\hat{h}}{\overline{D}\bar{\varepsilon}} \frac{H}{k \cos h(kL)} \sin h(kx_3), \tag{7.11.23}$$

$$\Delta p = -\frac{\mu^p p_0}{D^p} \frac{\hat{e}\hat{h}}{\overline{D}\bar{\varepsilon}} \frac{H}{k \cos h(kL)} \sin h(kx_3), \tag{7.11.24}$$

$$\Delta n = \frac{\mu^n n_0}{D^n} \frac{\hat{e}\hat{h}}{\overline{D}\bar{\varepsilon}} \frac{H}{k \cos h(kL)} \sin h(kx_3), \tag{7.11.25}$$

where

$$k^2 = \frac{qA^{(1)}}{\bar{\varepsilon}} \left(\frac{\mu^p}{D^p} p_0 + \frac{\mu^n}{D^n} n_0 \right), \quad \bar{\varepsilon} = \hat{\varepsilon} + \frac{\hat{e}^2}{\overline{D}}. \tag{7.11.26}$$

We are interested in the electrical response of the beam under H, in particular the redistribution of mobile charges. Specifically, we examine electrons. The behaviors of holes are similar. From Eq. (7.11.25), we calculate the redistributed electrons in the right half of the beam as

$$Q^e = \left| \int_0^L -q\Delta n(2c)b\,dx_3 \right| = \gamma H, \qquad (7.11.27)$$

where

$$\gamma = \frac{\left|\hat{e}\hat{h}\right|}{\overline{D}} \left[1 - \frac{1}{\cosh(kL)} \right]. \qquad (7.11.28)$$

Equation (7.11.28) looks reasonable. A larger $\hat{h}$ and a smaller $\overline{D}$ lead to a larger deformation under H. Then a larger $\hat{e}$ implies a larger piezoelectrically produced polarization which drives the mobile charges. γ describes the strength of the coupling effect of interest, i.e., the redistribution of charges under H. It depends on the material and geometric parameters. For numerical results, consider a $CoFe_2O_4/ZnO/CoFe_2O_4$ beam with $L = 0.6$ μm, $h = c = 0.05$ μm and $b = 0.2$ μm. Figure 7.23 shows that γ assumes a maximum for a certain value of $h/(h + c)$ when $h + c$ is fixed. This is because either the piezoelectric semiconductor layers or the PM layers cannot be too thin. Otherwise, there will be insufficient mobile charges

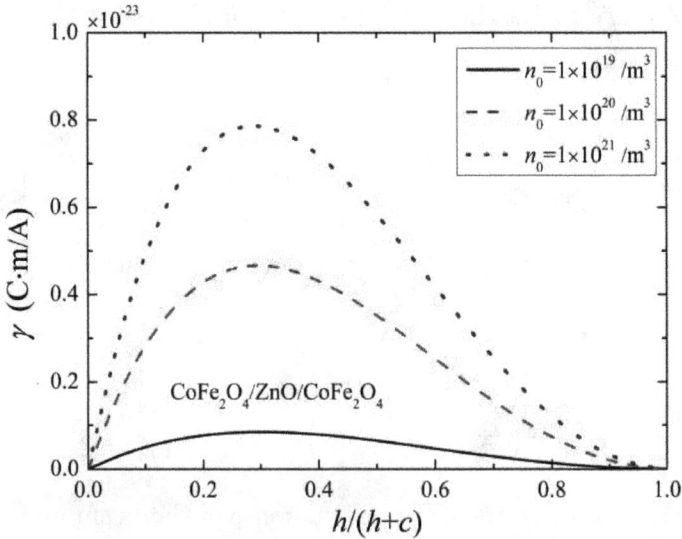

Fig. 7.23. γ versus $h/(h + c)$ while $h + c$ is fixed.

or insufficient piezomagnetically induced bending. The maximal γ is sensitive to the initial electron concentration described by n_0 which is determined by doping.

7.12 Piezomagnetic–Piezoelectric Semiconductor Unimorphs

Consider the thin beam in Fig. 7.24 consisting of a piezoelectric semiconductor (PS) layer (1) such as ZnO and a piezomagnetic (PM) dielectric layer (2) such as $CoFe_2O_4$ or Terfenol-D [55]. It may be called a unimorph because it has only one piezoelectric layer. Different from Secs. 7.7 and 7.8, in Fig. 7.24 the x_3-axis is at the geometric middle plane which experiences both axial extension and bending because of the lack of symmetry about the middle plane. The primed local coordinate system is for the PM layer only. The unimorph is under a transverse magnetic field H along the 3' direction of the PM layer. H is assumed known and uniform in the PM layer. It produces an axial extension in the PM layer through the piezomagnetic constant h_{31} in the local coordinate system and thus creates coupled extension and bending of the whole structure. The lateral surface of the unimorph is unelectroded. The electric field outside the unimorph is neglected.

For coupled extension and bending without shear deformation of the unimorph in the (x_3, x_2) plane, the extensional and bending

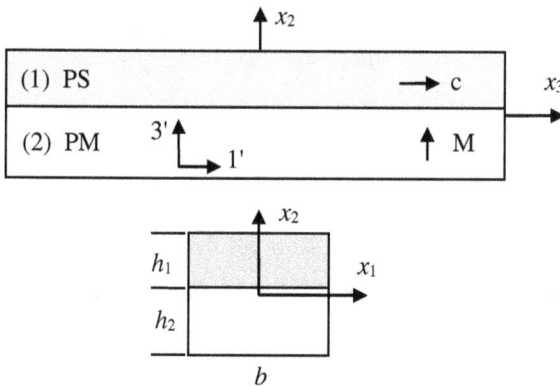

Fig. 7.24. A piezomagnetic–piezoelectric semiconductor unimorph.

displacements of the whole structure are described approximately by the following expressions:

$$u_2 \cong v(x_3, t),$$
$$u_3 \cong w(x_3, t) - x_2 v_{,3}, \qquad (7.12.1)$$

where v and w are the flexural and extensional displacements of the middle plane. Then the axial strain is expressed in terms of w and v as

$$S_3 = u_{3,3} = w_{,3} - x_2 v_{,33}. \qquad (7.12.2)$$

The electric potential and field in the entire unimorph as well as the carrier concentration perturbations in the PS layer only are approximated by

$$\varphi \cong \varphi(x_3, t), \quad E_1 = E_2 = 0, \quad E_3 = -\varphi_{,3}, \qquad (7.12.3)$$

$$\Delta p \cong \Delta p(x_3, t), \quad \Delta n \cong \Delta n(x_3, t). \qquad (7.12.4)$$

For the PS layer, under the usual stress relaxation for thin beams where the transverse normal stress components $T_1 = T_2 = 0$, the one-dimensional constitutive relations for the axial stress, electric displacement and current densities can be written as

$$T = \bar{c}^{(1)} S - \bar{e}^{(1)} E = \bar{c}^{(1)}(w_{,3} - x_2 v_{,33}) + \bar{e}^{(1)} \varphi_{,3},$$
$$D = \bar{e}^{(1)} S + \bar{\varepsilon}^{(1)} E = \bar{e}^{(1)}(w_{,3} - x_2 v_{,33}) - \bar{\varepsilon}^{(1)} \varphi_{,3}, \qquad (7.12.5)$$

$$J^p \cong q p_0 \mu^p E - q D^p \frac{\partial(\Delta p)}{\partial z},$$
$$J^n \cong q n_0 \mu^n E + q D^n \frac{\partial(\Delta n)}{\partial z}, \qquad (7.12.6)$$

where

$$S = S_3, \quad T = T_3, \quad E = E_3, \quad D = D_3,$$
$$J^p = J_3^p, \quad J^n = J_3^n, \qquad (7.12.7)$$

$$\mu^p = \mu_{33}^p, \quad D^p = D_{33}^p,$$
$$\mu^n = \mu_{33}^n, \quad D^n = D_{33}^n. \qquad (7.12.8)$$

The effective one-dimensional material constants under $T_1 = T_2 = 0$ are related to the three-dimensional material constants through

$$\bar{c}^{(1)} = c_{33}^{(1)} - \frac{2(c_{13}^{(1)})^2}{c_{11}^{(1)} + c_{12}^{(1)}}, \quad \bar{e}^{(1)} = e_{33}^{(1)} - \frac{2c_{13}^{(1)} e_{31}^{(1)}}{c_{11}^{(1)} + c_{12}^{(1)}},$$

$$\bar{\varepsilon}^{(1)} = \varepsilon_{33}^{(1)} + \frac{2(e_{31}^{(1)})^2}{c_{11}^{(1)} + c_{12}^{(1)}}.$$

(7.12.9)

For the PM layer, under the stress relaxation for thin beams where in the local coordinate system we have $T_2' = T_3' = 0$, the one-dimensional constitutive relation for the axial fields takes the following form:

$$T = \bar{c}^{(2)} S - \bar{h}^{(2)} H = \bar{c}^{(2)}(w_{,3} - x_2 v_{,33}) - \bar{h}^{(2)} H,$$

$$D = \bar{\varepsilon}^{(2)} E,$$

(7.12.10)

where the effective one-dimensional material constants have the following expressions:

$$\bar{c}^{(2)} = c_{11}^{(2)} + \frac{2c_{12}^{(2)}(c_{13}^{(2)})^2 - c_{33}^{(2)}(c_{12}^{(2)})^2 - c_{11}^{(2)}(c_{13}^{(2)})^2}{c_{11}^{(2)} c_{33}^{(2)} - (c_{13}^{(2)})^2},$$

$$\bar{h}^{(2)} = h_{31}^{(2)}$$

$$- \frac{c_{12}^{(2)} c_{33}^{(2)} h_{31}^{(2)} - c_{12}^{(2)} c_{13}^{(2)} h_{33}^{(2)} + c_{13}^{(2)} c_{11}^{(2)} h_{33}^{(2)} - (c_{13}^{(2)})^2 h_{31}^{(2)}}{c_{11}^{(2)} c_{33}^{(2)} - (c_{13}^{(2)})^2},$$

$$\bar{\varepsilon}^{(2)} = \varepsilon_{11}^{(2)}.$$

(7.12.11)

For the two layers together, the total axial force N, transverse shear force Q, bending moment M and axial electric displacement $\hat{D}$ are defined by the following integrations over a cross-section:

$$N = \int_A T \, dA = \hat{c} w_{,3} - \hat{c}' v_{,33} + \hat{e} \varphi_{,3} - \hat{h} H,$$

(7.12.12)

$$Q = \int_A T_{32} \, dA,$$

(7.12.13)

$$M = \int_A Tx_2 \, dA = \hat{c}'w_{,3} - \overline{D}v_{,33} + \hat{e}\frac{h_2}{2}\varphi_{,3} + \hat{h}\frac{h_1}{2}H, \quad (7.12.14)$$

$$\hat{D} = \int_A D \, dA = \hat{e}w_{,3} - \hat{e}\frac{h_2}{2}v_{,33} - \hat{\varepsilon}\varphi_{,3}, \quad (7.12.15)$$

where

$$\hat{c} = \overline{c}^{(1)}A^{(1)} + \overline{c}^{(2)}A^{(2)}, \quad \hat{c}' = \left(\overline{c}^{(1)} - \overline{c}^{(2)}\right)\frac{1}{2}bh_1h_2,$$

$$\hat{e} = \overline{e}^{(1)}A^{(1)}, \quad \hat{h} = \overline{h}^{(2)}A^{(2)},$$

$$A^{(1)} = bh_1, \quad A^{(2)} = bh_2,$$

$$\overline{D} = \overline{c}^{(1)}\frac{b}{24}\left[(h_1 + h_2)^3 - (h_2 - h_1)^3\right]$$

$$+ \overline{c}^{(2)}\frac{b}{24}\left[(h_1 + h_2)^3 + (h_2 - h_1)^3\right],$$

$$\hat{\varepsilon} = \overline{\varepsilon}^{(1)}A^{(1)} + \overline{\varepsilon}^{(2)}A^{(2)}.$$

The constitutive relation for the shear force Q is not derived from Eq. (7.12.13). It will be provided by the following shear force-bending moment relation.

From the equations of motion of the differential element of the unimorph in Fig. 7.25 in the x_3 and x_2 directions as well as its moment equation, we obtain

$$N_{,3} + F_3 = b\left(\rho^{(1)}h_1 + \rho^{(2)}h_2\right)\ddot{w} + b\left(\rho^{(2)} - \rho^{(1)}\right)\frac{h_1h_2}{2}\ddot{v}_{,3},$$

$$Q_{,3} + F_2 = b\left(\rho^{(1)}h_1 + \rho^{(2)}h_2\right)\ddot{v}, \quad (7.12.18)$$

$$M_{,3} - Q = 0,$$

where the rotatory inertia has been neglected. The second term on the right-hand side of Eq. $(7.12.8)_1$ is due to the fact that the center of mass of the element in Fig. 7.25 deviates a little from the x_3-axis at the geometric middle plane. Equation $(7.12.18)_3$ is the shear force-bending moment relation. Equations $(7.12.18)_3$ and (7.12.14) together imply the following constitutive relation for the shear force:

$$Q = M_{,3} = \hat{c}'w_{,33} - \overline{D}v_{,333} + \hat{e}\frac{h_2}{2}\varphi_{,33}. \quad (7.12.19)$$

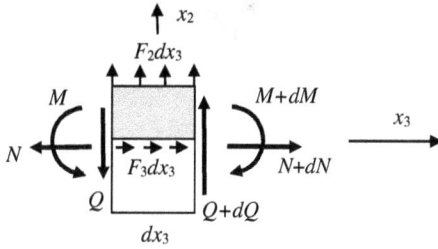

Fig. 7.25. A differential element of the unimorph under mechanical loads.

Similar to the derivation of Eq. (7.12.18), by considering the differential element in Fig. 7.25 under electrical loads, the charge equation of electrostatic can be written as

$$\hat{D}_{,3} = q(\Delta p - \Delta n)A^{(1)}. \tag{7.12.20}$$

The one-dimensional conservation of holes and electrons can be simply reduced from the three-dimensional equations as

$$q\frac{\partial}{\partial t}(\Delta p) = -J^p_{,3},$$
$$q\frac{\partial}{\partial t}(\Delta n) = J^n_{,3}. \tag{7.12.21}$$

The substitution of Eqs. (7.12.12), (7.12.15), (7.12.19) and (7.12.6) into Eqs. (7.12.18)$_{1,2}$, (7.12.20) and (7.12.21) gives five equations for w, v, φ, Δp and Δn.

Appendix 1

List of Symbols

Because of the multi-physical fields involved and the use of one- and three-dimensional theories, some of the symbols may have different meanings in different chapters or sections.

ρ — mass density, radius of curvature
E — Young's modulus
ν — Poisson's ratio
G — shear modulus
N — extensional force in a rod
M — torque, twisting or bending moment
Q — transverse shear force in a beam
F_i — distributed force per unit length of a beam
m_i — distributed moment per unit length of a beam
ψ — angle of twist, shear displacement, magnetic potential
A — cross-sectional area of a rod
$I, \bar{I}$ — moments of inertia, currents
$I_p, \bar{I}_p$ — polar moments of inertia
$\bar{D}$ — bending stiffness of a beam
λ — Timoshenko's shear correction factor for beam bending
i — imaginary unit
i, j, k, l — three-dimensional tensor indices. Range: 1–3
p, q — three-dimensional matrix indices. Range: 1–6
x_k — Cartesian coordinates
i, j, k — basis vectors of Cartesian coordinates
$\mathbf{e}_1, \mathbf{e}_2, \mathbf{e}_3$ — basis vectors of Cartesian coordinates

$\mathbf{e}_r$, $\mathbf{e}_\theta$, $\mathbf{e}_z$ — basis vectors of cylindrical coordinates

δ_{ij} — Kronecker delta

f_j — distributed body force per unit volume

u_k — mechanical displacement vector

T_{kl} — stress tensor

S_{kl} — strain tensor

c_{ijkl}, s_{ijkl} — elastic stiffness and compliance

U — internal energy density per unit volume

U^F — electromagnetic field energy density per unit volume

$\hat{U}$ — sum of field and internal energy densities ($= U^F + U$)

F — Helmholtz free energy density per unit volume

H — enthalpy density per unit volume

G — Gibbs free energy density per unit volume

T — absolute temperature

Θ_0 — uniform reference temperature

θ — small temperature deviation from Θ_0

C_V — specific heat per unit volume

η — entropy density per unit volume

h_i — heat flux vector

r — heat source per unit volume

κ_{kl} — heat conduction coefficients

λ_{kl} — thermoelastic constants

α_{ij} — coefficients of thermal expansion

ε_0 — vacuum electric permittivity

μ_0 — vacuum magnetic permeability

q — elementary charge

φ — electrostatic potential

E_i — electric field

P_i — electric polarization

D_i — electric displacement

B_i — magnetic flux or induction

M_i — magnetization

H_i — magnetic field

S_i — Poynting vector

ρ^t — total charge density per unit volume

ρ^P — effective polarization charge density per unit volume

σ^P — effective polarization charge density per unit area

ρ^e — difference of total and polarization charges ($\rho^t - \rho^P$)

$\mathbf{J}^t$ — total current density

$\mathbf{J}^M$ — effective magnetization current density

$\mathbf{J}$ — difference of total and magnetization currents ($\mathbf{J}^t - \mathbf{J}^M$)

χ_{ij}^e, χ_{ij}^M — electric and magnetic susceptibility

ε_{ij}^S, ε_{ij}^T — dielectric constants

μ_{ij} — magnetic permeability

p, n — concentrations of holes and electrons

$p_0 = N_A^-$, $n_0 = N_D^+$ — concentrations of ionized acceptors and donors

Δp, Δn — small carrier concentration perturbations from p_0 and n_0

γ^p, γ^n — body sources of holes and electrons

k_B — Boltzmann constant

$\mathbf{J}^p$, $\mathbf{J}^n$ — current densities of holes and electrons

μ_{ij}^p, μ_{ij}^n — mobility of holes and electrons

D_{ij}^p, D_{ij}^n — diffusion constants of holes and electrons

λ_D — Debye–Hückel length

m_{ij} — magnetoelectric constants

p_j — pyroelectric constants

q_j — pyromagnetic constants

e_{ijk}, d_{ijk} — piezoelectric constants

h_{ijk} — piezomagnetic constants

k^2 — electromechanical coupling factor ($= e^2/(\varepsilon c)$)

c' — piezoelectrically stiffened elastic constant

$\hat{D}$ — total axial electric displacement in a rod

$\bar{c}$ — effective one-dimensional elastic constant

$\bar{e}$ — effective one-dimensional piezoelectric constant

$\bar{h}$ — effective one-dimensional piezomagnetic constant

$\bar{\varepsilon}$ — effective one-dimensional dielectric constant

$\bar{\lambda}$ — effective one-dimensional thermoelastic constant

$\bar{p}$ — effective one-dimensional pyroelectric constant

κ — Mindlin's shear correction factor for bending

Appendix 2

Material Constants

Vacuum electric permittivity $\varepsilon_0 = 8.854 \times 10^{-12}\,\mathrm{F/m}$
Vacuum magnetic permeability $\mu_0 = 12.57 \times 10^{-7}\,\mathrm{H/m}$
Elementary charge $q = 1.602 \times 10^{-19}\,\mathrm{C}$
Boltzmann constant $k_B = 1.381 \times 10^{-23}\,\mathrm{J/K}$
$k_B T/q = 0.0259\,\mathrm{V}$ at room temperature $300\,\mathrm{K}$ [18]

Aluminum nitride (AlN) [56]

$\rho = 3,260\,\mathrm{kg/m}^3$

$$[c_{pq}] = \begin{bmatrix} 345 & 125 & 120 & 0 & 0 & 0 \\ 125 & 345 & 120 & 0 & 0 & 0 \\ 120 & 120 & 395 & 0 & 0 & 0 \\ 0 & 0 & 0 & 118 & 0 & 0 \\ 0 & 0 & 0 & 0 & 118 & 0 \\ 0 & 0 & 0 & 0 & 0 & 110 \end{bmatrix} \times 10^9 \mathrm{N/m}^2$$

$$[e_{ip}] = \begin{bmatrix} 0 & 0 & 0 & 0 & -0.48 & 0 \\ 0 & 0 & 0 & -0.48 & 0 & 0 \\ -0.58 & -0.58 & 1.55 & 0 & 0 & 0 \end{bmatrix} \mathrm{C/m}^2$$

$$[\varepsilon_{ij}] = \begin{bmatrix} 8.0 & 0 & 0 \\ 0 & 8.0 & 0 \\ 0 & 0 & 9.5 \end{bmatrix} \times 10^{-11}\,\mathrm{F/m}$$

Barium titanate (BaTiO$_3$) [57]

$\rho = 5,800 \, \text{kg/m}^3$

$$[c_{pq}] = \begin{bmatrix} 166.0 & 77.0 & 78.0 & 0 & 0 & 0 \\ 77.0 & 166.0 & 78.0 & 0 & 0 & 0 \\ 78.0 & 78.0 & 162.0 & 0 & 0 & 0 \\ 0 & 0 & 0 & 43.0 & 0 & 0 \\ 0 & 0 & 0 & 0 & 43.0 & 0 \\ 0 & 0 & 0 & 0 & 0 & 44.5 \end{bmatrix} \times 10^9 \, \text{N/m}^2$$

$$[e_{jq}] = \begin{bmatrix} 0 & 0 & 0 & 0 & 11.6 & 0 \\ 0 & 0 & 0 & 11.6 & 0 & 0 \\ -4.4 & -4.4 & 18.6 & 0 & 0 & 0 \end{bmatrix} \text{C/m}^2$$

$$[\varepsilon_{ij}] = \begin{bmatrix} 11.2 & 0 & 0 \\ 0 & 11.2 & 0 \\ 0 & 0 & 12.6 \end{bmatrix} \times 10^{-9} \, \text{F/m}$$

Cadmium selenide (CdSe) [24]

$\rho = 4,820 \, \text{kg/m}^3$

$$[c_{pq}] = \begin{bmatrix} 90.7 & 58.1 & 51.0 & 0 & 0 & 0 \\ 58.1 & 90.7 & 51.0 & 0 & 0 & 0 \\ 51.0 & 51.0 & 93.8 & 0 & 0 & 0 \\ 0 & 0 & 0 & 15.04 & 0 & 0 \\ 0 & 0 & 0 & 0 & 15.04 & 0 \\ 0 & 0 & 0 & 0 & 0 & 16.3 \end{bmatrix} \times 10^9 \, \text{N/m}^2$$

$$[e_{ip}] = \begin{bmatrix} 0 & 0 & 0 & 0 & -0.21 & 0 \\ 0 & 0 & 0 & -0.21 & 0 & 0 \\ -0.24 & -0.24 & 0.44 & 0 & 0 & 0 \end{bmatrix} \text{C/m}^2$$

$$[\varepsilon_{ij}] = \begin{bmatrix} 9.02 & 0 & 0 \\ 0 & 9.02 & 0 \\ 0 & 0 & 9.53 \end{bmatrix} \varepsilon_0$$

Gallium arsenide (GaAs) [24,18,19]

$\rho = 5,307 \, \text{kg/m}^2$

$$[c_{pq}] = \begin{bmatrix} 11.88 & 5.38 & 5.38 & 0 & 0 & 0 \\ 5.38 & 11.88 & 5.38 & 0 & 0 & 0 \\ 5.38 & 5.38 & 11.88 & 0 & 0 & 0 \\ 0 & 0 & 0 & 5.94 & 0 & 0 \\ 0 & 0 & 0 & 0 & 5.94 & 0 \\ 0 & 0 & 0 & 0 & 0 & 5.94 \end{bmatrix} \times 10^{10} \text{N/m}^2$$

$$[e_{ip}] = \begin{bmatrix} 0 & 0 & 0 & 0.154 & 0 & 0 \\ 0 & 0 & 0 & 0 & 0.154 & 0 \\ 0 & 0 & 0 & 0 & 0 & 0.154 \end{bmatrix} \text{C/m}^2$$

$$[\varepsilon_{ij}] = \begin{bmatrix} 12.5 & 0 & 0 \\ 0 & 12.5 & 0 \\ 0 & 0 & 12.5 \end{bmatrix} \times 8.85 \times 10^{-12} \text{F/m}$$

$\mu^n = 8500 \, \text{cm}^2/\text{V} \cdot \text{s}$

$\mu^p = 400 \, \text{cm}^2/\text{V} \cdot \text{s}$

Germanium (Ge) [24,19]

$\rho = 5,327 \, \text{kg/m}^3$

$$[c_{pq}] = \begin{bmatrix} 128.9 & 48.3 & 48.3 & 0 & 0 & 0 \\ 48.3 & 128.9 & 48.3 & 0 & 0 & 0 \\ 48.3 & 48.3 & 128.9 & 0 & 0 & 0 \\ 0 & 0 & 0 & 67.1 & 0 & 0 \\ 0 & 0 & 0 & 0 & 67.1 & 0 \\ 0 & 0 & 0 & 0 & 0 & 67.1 \end{bmatrix} \times 10^9 \text{N/m}^2$$

$$[\varepsilon_{ij}] = \begin{bmatrix} 0.1398932 & 0 & 0 \\ 0 & 0.1398932 & 0 \\ 0 & 0 & 0.1398932 \end{bmatrix} \times 10^{-9} \text{F/m}$$

$\mu^n = 3900 \, \text{cm}^2/\text{V} \cdot \text{s}$

$\mu^p = 1900 \, \text{cm}^2/\text{V} \cdot \text{s}$

Lithium niobate (LiNbO₃) [58,22]

$\rho = 4,700\,\mathrm{kg/m}^3$

$$[c_{pq}] = \begin{bmatrix} 2.03 & 0.53 & 0.75 & 0.09 & 0 & 0 \\ 0.53 & 2.03 & 0.75 & -0.09 & 0 & 0 \\ 0.75 & 0.75 & 2.45 & 0 & 0 & 0 \\ 0.09 & -0.09 & 0 & 0.60 & 0 & 0 \\ 0 & 0 & 0 & 0 & 0.60 & 0.09 \\ 0 & 0 & 0 & 0 & 0.09 & 0.75 \end{bmatrix} \times 10^{11}\,\mathrm{N/m}^2$$

$$[e_{ip}] = \begin{bmatrix} 0 & 0 & 0 & 0 & 3.70 & -2.50 \\ -2.50 & 2.50 & 0 & 3.70 & 0 & 0 \\ 0.20 & 0.20 & 1.30 & 0 & 0 & 0 \end{bmatrix} \mathrm{C/m}^2$$

$$[\varepsilon_{ij}] = \begin{bmatrix} 38.9 & 0 & 0 \\ 0 & 38.9 & 0 \\ 0 & 0 & 25.7 \end{bmatrix} \times 10^{-11}\,\mathrm{F/m}$$

Lithium tantalate (LiTaO₃) [58,22]

$\rho = 7,450\,\mathrm{kg/m}^3$,

$$[c_{pq}] = \begin{bmatrix} 2.33 & 0.47 & 0.80 & -0.11 & 0 & 0 \\ 0.47 & 2.33 & 0.80 & 0.11 & 0 & 0 \\ 0.80 & 0.80 & 2.75 & 0 & 0 & 0 \\ -0.11 & -0.11 & 0 & 0.94 & 0 & 0 \\ 0 & 0 & 0 & 0 & 0.94 & -0.11 \\ 0 & 0 & 0 & 0 & -0.11 & 0.93 \end{bmatrix} \times 10^{11}\,\mathrm{N/m}^2$$

$$[e_{ip}] = \begin{bmatrix} 0 & 0 & 0 & 0 & 2.6 & -1.6 \\ -1.6 & 1.6 & 0 & 2.6 & 0 & 0 \\ 0 & 0 & 1.9 & 0 & 0 & 0 \end{bmatrix} \mathrm{C/m}^2$$

$$[\varepsilon_{ij}] = \begin{bmatrix} 36.3 & 0 & 0 \\ 0 & 36.3 & 0 \\ 0 & 0 & 38.2 \end{bmatrix} \times 10^{-11} \text{ F/m}$$

PZT-2 [24]

$\rho = 7,600 \text{ kg/m}^3$

$$[c_{pq}] = \begin{bmatrix} 13.5 & 6.79 & 6.81 & 0 & 0 & 0 \\ 6.79 & 13.5 & 6.81 & 0 & 0 & 0 \\ 6.81 & 6.81 & 11.3 & 0 & 0 & 0 \\ 0 & 0 & 0 & 2.22 & 0 & 0 \\ 0 & 0 & 0 & 0 & 2.22 & 0 \\ 0 & 0 & 0 & 0 & 0 & 3.36 \end{bmatrix} \times 10^{10} \text{ N/m}^2$$

$$[e_{ip}] = \begin{bmatrix} 0 & 0 & 0 & 0 & 9.8 & 0 \\ 0 & 0 & 0 & 9.8 & 0 & 0 \\ -1.9 & -1.9 & 9.0 & 0 & 0 & 0 \end{bmatrix} \text{ C/m}^2$$

$$[\varepsilon_{ij}] = \begin{bmatrix} 504\varepsilon_0 & 0 & 0 \\ 0 & 504\varepsilon_0 & 0 \\ 0 & 0 & 260\varepsilon_0 \end{bmatrix}$$

PZT-4 [57]

$\rho = 7,600 \text{ kg/m}^3$

$$[c_{pq}] = \begin{bmatrix} 138.5 & 77.37 & 73.64 & 0 & 0 & 0 \\ 77.37 & 138.5 & 73.64 & 0 & 0 & 0 \\ 73.64 & 73.64 & 114.8 & 0 & 0 & 0 \\ 0 & 0 & 0 & 25.6 & 0 & 0 \\ 0 & 0 & 0 & 0 & 25.6 & 0 \\ 0 & 0 & 0 & 0 & 0 & 30.6 \end{bmatrix} \times 10^9 \text{ N/m}^2$$

$$[e_{jq}] = \begin{bmatrix} 0 & 0 & 0 & 0 & 12.72 & 0 \\ 0 & 0 & 0 & 12.72 & 0 & 0 \\ -5.2 & -5.2 & 15.08 & 0 & 0 & 0 \end{bmatrix} \text{C/m}^2$$

$$[\varepsilon_{ij}] = \begin{bmatrix} 13.06 & 0 & 0 \\ 0 & 13.06 & 0 \\ 0 & 0 & 11.15 \end{bmatrix} \times 10^{-9} \text{ F/m}$$

PZT-5A [57]

$\rho = 7,750 \, \text{kg/m}^3$

$$[c_{pq}] = \begin{bmatrix} 99.201 & 54.016 & 50.778 & 0 & 0 & 0 \\ 54.016 & 99.201 & 50.778 & 0 & 0 & 0 \\ 50.788 & 50.788 & 86.856 & 0 & 0 & 0 \\ 0 & 0 & 0 & 21.1 & 0 & 0 \\ 0 & 0 & 0 & 0 & 21.1 & 0 \\ 0 & 0 & 0 & 0 & 0 & 22.6 \end{bmatrix} \times 10^9 \text{ N/m}^2$$

$$[e_{jq}] = \begin{bmatrix} 0 & 0 & 0 & 0 & 12.322 & 0 \\ 0 & 0 & 0 & 12.322 & 0 & 0 \\ -7.209 & -7.209 & 15.118 & 0 & 0 & 0 \end{bmatrix} \text{C/m}^2$$

$$[\varepsilon_{ij}] = \begin{bmatrix} 15.3 & 0 & 0 \\ 0 & 15.3 & 0 \\ 0 & 0 & 15.0 \end{bmatrix} \times 10^{-9} \text{ F/m}$$

PZT-5H [24]

$\rho = 7,500 \, \text{kg/m}^3$

$$[c_{pq}] = \begin{bmatrix} 12.6 & 7.95 & 8.41 & 0 & 0 & 0 \\ 7.95 & 12.6 & 8.41 & 0 & 0 & 0 \\ 8.41 & 8.41 & 11.7 & 0 & 0 & 0 \\ 0 & 0 & 0 & 2.30 & 0 & 0 \\ 0 & 0 & 0 & 0 & 2.30 & 0 \\ 0 & 0 & 0 & 0 & 0 & 2.33 \end{bmatrix} \times 10^{10} \text{ N/m}^2$$

$$[e_{ip}] = \begin{bmatrix} 0 & 0 & 0 & 0 & 17.0 & 0 \\ 0 & 0 & 0 & 17.0 & 0 & 0 \\ -6.5 & -6.5 & 23.3 & 0 & 0 & 0 \end{bmatrix} C/m^2$$

$$[\varepsilon_{ij}] = \begin{bmatrix} 1700\varepsilon_0 & 0 & 0 \\ 0 & 1700\varepsilon_0 & 0 \\ 0 & 0 & 1470\varepsilon_0 \end{bmatrix}$$

Silicon (Si) [24,19]

$\rho = 2,332 \text{ kg/m}^3$

$$[c_{pq}] = \begin{bmatrix} 16.57 & 6.39 & 6.39 & 0 & 0 & 0 \\ 6.39 & 16.57 & 6.39 & 0 & 0 & 0 \\ 6.39 & 6.39 & 16.57 & 0 & 0 & 0 \\ 0 & 0 & 0 & 7.956 & 0 & 0 \\ 0 & 0 & 0 & 0 & 7.956 & 0 \\ 0 & 0 & 0 & 0 & 0 & 7.956 \end{bmatrix} \times 10^{10} N/m^2$$

$$[\varepsilon_{ij}] = \begin{bmatrix} 11.7\varepsilon_0 & 0 & 0 \\ 0 & 11.7\varepsilon_0 & 0 \\ 0 & 0 & 11.7\varepsilon_0 \end{bmatrix}$$

$$\mu^n = 1500 \text{ cm}^2/\text{V} \cdot \text{s}$$
$$\mu^p = 450 \text{ cm}^2/\text{V} \cdot \text{s}$$

Zinc oxide (ZnO) [24,19]

$\rho = 5,680 \text{ kg/m}^3$

$$[c_{pq}] = \begin{bmatrix} 20.97 & 12.11 & 10.51 & 0 & 0 & 0 \\ 12.11 & 20.97 & 10.51 & 0 & 0 & 0 \\ 10.51 & 10.51 & 21.09 & 0 & 0 & 0 \\ 0 & 0 & 0 & 4.247 & 0 & 0 \\ 0 & 0 & 0 & 0 & 4.247 & 0 \\ 0 & 0 & 0 & 0 & 0 & 4.43 \end{bmatrix} \times 10^{10} \text{ N/m}^2$$

$$[e_{ip}] = \begin{bmatrix} 0 & 0 & 0 & 0 & -0.48 & 0 \\ 0 & 0 & 0 & -0.48 & 0 & 0 \\ -0.573 & -0.573 & 1.32 & 0 & 0 & 0 \end{bmatrix} \text{C/m}^2$$

$$[\varepsilon_{ij}] = \begin{bmatrix} 8.55\varepsilon_0 & 0 & 0 \\ 0 & 8.55\varepsilon_0 & 0 \\ 0 & 0 & 10.2\varepsilon_0 \end{bmatrix}$$

$$\mu^n = 200 \text{ cm}^2/\text{V} \cdot \text{s}$$
$$\mu^p = 180 \text{ cm}^2/\text{V} \cdot \text{s}$$

References

[1] J.M. Gere and S.P. Timoshenko, *Mechanics of Materials*, 2nd ed., Wadsworth, Belmont, California, 1984.

[2] L. Meirovitch, *Analytical Methods in Vibrations*, Macmillan, London, 1967.

[3] K.F. Graff, *Wave Motion in Elastic Solids*, Dover, New York, 1991.

[4] S.P. Timoshenko and J.N. Goodier, *Theory of Elasticity*, McGraw-Hill, New York, 1970.

[5] R.D. Mindlin, *An Introduction to the Mathematical Theory of Vibrations of Elastic Plates*, J.S. Yang (ed.), World Scientific, Singapore, 2006.

[6] S.P. Timoshenko, On the correction for shear of the differential equation for transverse vibrations of prismatic bars, *Lond. Edinb. Dublin Philos. Mag. J. Sci.*, 41, 744–746, 1921.

[7] R.D. Mindlin, Low frequency vibrations of elastic bars, *Int. J. Solids Struct.*, 12, 27–49, 1976.

[8] H. Parkus, *Thermoelasticity*, Blaisdell, Waltham, Massachusetts, 1968.

[9] B.A. Boley and J.H. Weiner, *Theory of Thermal Stresses*, John Willey & Sons, New York, 1960.

[10] W. Nowacki, *Thermoelasticity*, PWN, Warszawa, 1962.

[11] R.R. Cheng, C.L. Zhang, W.Q. Chen and J.S. Yang, Temperature effects on mobile charges in extension of composite fibers of piezoelectric dielectrics and nonpiezoelectric semiconductors, *Int. J. Appl. Mech.*, 11, 1950088, 2019.

[12] S. Ju, J.S. Yang and H.F. Zhang, Effects of mobile charges on interface thermal stresses in a piezoelectric-semiconductor composite rod, *Arch. Appl. Mech.*, 92, 1633–1641, 2022.

[13] A.T. Adams, *Electromagnetics for Engineers*, Ronald, 1971.

[14] W.K.H. Panofsky and M. Phillips, *Classical Electricity and Magnetism*, Addison Willey, Reading, Massachusetts, 1962.

[15] D.J. Jackson, *Classical Electrodynamics*, 2nd ed., John Willey & Sons, Singapore, 1990.

[16] L.D. Landau and E.M. Lifshitz, *Electrodynamics of Continuous Media*, 2nd ed., Butterworth-Heinemann, Linacre House, Jordan Hill, Oxford, 1984.

[17] H.F. Tiersten, *A Development of the Equations of Electromagnetism in Material Continua*, Springer, New York, 1990.

[18] R.F. Pierret, *Semiconductor Device Fundamentals*, Pearson, Uttar Pradesh, India, 1996.

[19] S.M. Sze, *Physics of Semiconductor Devices*, John Wiley & Sons, New York, 1981.

[20] S. Selberherr, *Analysis and Simulation of Semiconductor Devices*, Springer-Verlag, New York, 1984.

[21] Y.X. Luo, C.L. Zhang, W.Q. Chen and J.S. Yang, An analysis of PN junctions in piezoelectric semiconductors, *J. Appl. Phys.*, 122, 204502, 2017.

[22] H.F. Tiersten, *Linear Piezoelectric Plate Vibrations*, Plenum, New York, 1969.

[23] A.H. Meitzler, D. Berlincourt, F.S. Welsh III, H.F. Tiersten, G.A. Coquin and A.W. Warner, *IEEE Standard on Piezoelectricity*, IEEE, New York, 1988.

[24] B.A. Auld, *Acoustic Fields and Waves in Solids*, vol. 1, Wiley, New York, 1973.

[25] J.S. Yang, *An Introduction to the Theory of Piezoelectricity*, 2nd edn., Springer, New York, 2018.

[26] R.D. Mindlin, On the equations of motion of piezoelectric crystals, in: *Problems of Continuum Mechanics*, J.R.M. Radok (ed.), Soc. Ind. Appl. Math., Philadelphia, 1961, pp. 282–290.

[27] R.D. Mindlin, Equations of high frequency vibrations of thermopiezo-electric crystal plates, *Int. J. Solids Struct.*, 10, 625–637, 1974.

[28] A.C. Eringen and G.A. Maugin, *Electrodynamics of Continua*, vol. I, Springer-Verlag, New York, 1990.

[29] H.G. de Lorenzi and H.F. Tiersten, On the interaction of the electromagnetic field with heat conducting deformable semiconductors, *J. Math. Phys.*, 16, 938–957, 1975.

[30] G.A. Maugin and N. Daher, Phenomenological theory of elastic semiconductors, *Int. J. Engng Sci.*, 24, 703–731, 1986.

[31] Z.L. Wang, *Piezotronics and Piezo-Phototronics*, Springer-Verlag, Berlin, Heidelberg, 2012.

[32] J.S. Yang, *Analysis of Piezoelectric Semiconductor Structures*, Springer Nature, Switzerland, 2020.

[33] D.L. White, Amplification of ultrasonic waves in piezoelectric semiconductors, *J. Appl. Phys.*, 33, 2547–2554, 1962.

[34] J.S. Yang, *Mechanics of Piezoelectric Structures*, 2nd ed., World Scientific, Singapore, 2020.

[35] J.S. Yang and H.Y. Fang, A piezoelectric gyroscope based on extensional vibrations of rods, *Int. J. Appl. Electrom.*, 17, 289–300, 2003.

[36] C.L. Zhang, X.Y. Wang, W.Q. Chen and J.S. Yang, An analysis of the extension of a ZnO piezoelectric semiconductor nanofiber under an axial force, *Smart Mater. Struct.*, 26, 025030, 2017.

[37] C. Liang, C.L. Zhang, W.Q. Chen and J.S. Yang, Electrical response of a multiferroic composite semiconductor fiber under a local magnetic field, *Acta Mech. Solida Sin.*, 33, 663–673, 2020.

[38] G.L. Wang, X.L. Liu, W.J. Feng and J.S. Yang, Magnetically induced carrier distribution and extension in a composite rod of piezoelectric semiconductors and piezomagnetics, *Materials*, 13, 3115, 2020.

[39] S.Y. Lin, Torsional vibration of coaxially segmented, tangentially polarized piezoelectric ceramic tubes, *J. Acoust. Soc. Am.*, 99, 3476–3480, 1996.

[40] Y.T. Guo, C.L. Zhang, W.Q. Chen and J.S. Yang, Interaction between torsional deformation and mobile charges in a composite rod of piezoelectric dielectrics and nonpiezoelectric semiconductors, *Mech. Adv. Mater. Struct.*, 29, 1449–1455, 2022.

[41] R.D. Mindlin, Low frequency vibrations of elastic bars, *Int. J. Solids Struct.*, 12, 27–49, 1976.

[42] M.C. Dokmeci, A theory of high frequency vibrations of piezoelectric crystal bars, *Int. J. Solids Struct.*, 10, 401–409, 1974.

[43] C.S. Chou, J.W. Yang, Y.C. Huang and H.J. Yang, Analysis on vibrating piezoelectric beam gyroscope, *Int. J. Appl. Electrom.*, 2, 227–241, 1991.

[44] J.S. Yang, Equations for the extension and flexure of a piezoelectric beam with rectangular cross section and applications, *Int. J. Appl. Electrom.*, 9, 409–420, 1998.

[45] P. Li, F. Jin and J. Ma, One-dimensional dynamic equations of a piezoelectric semiconductor beam with a rectangular cross section and their application in static and dynamic characteristic analysis, *Appl. Math. Mech.*, 39, 685–702, 2018.

[46] Y.L. Qu, F. Jin and J.S. Yang, Torsion of a piezoelectric semiconductor rod of cubic crystals with consideration of warping and in-plane shear of its rectangular cross section, *Mech. Mater.*, 172, 104407, 2022.

[47] J.L. Bleustein and R. Stanley, A dynamical theory of torsion, *Int. J. Solids Struct.*, 6, 569–586, 1970.

[48] C.L. Zhang, X.Y. Wang, W.Q. Chen and J.S. Yang, Bending of a cantilever piezoelectric semiconductor fiber under an end force, in: *Generalized Models and Non-Classical Approaches in Complex Materials*, H. Altenbach, *et al.* (ed.), Springer, Cham, Switzerland, 2018, pp. 261–278.

[49] J.S. Yang, H.Y. Fang and Q. Jiang, Analysis of a ceramic bimorph piezoelectric gyroscope, *Int. J. Appl. Electrom.*, 10, 459–473, 1999.

[50] S.N. Jiang, X.F. Li, S.H. Guo, Y.T. Hu, J.S. Yang and Q. Jiang, Performance of a piezoelectric bimorph for scavenging vibration energy, *Smart Mater. Struct.*, 14, 769–714, 2005.

[51] Y.X. Luo, C.L. Zhang, W.Q. Chen and J.S. Yang, Piezopotential in a bended composite fiber made of a semiconductive core and of two piezoelectric layers with opposite polarities, *Nano Energy*, 54, 341–348, 2018.

[52] Y.X. Luo, C.L. Zhang, W.Q. Chen and J.S. Yang, Thermally induced electromechanical fields in unimorphs of piezoelectric dielectrics and nonpiezoelectric semiconductors, *Integr. Ferroelectr.*, 211, 117–131, 2020.

[53] K. Fang, Z.H. Qian and J.S. Yang, Piezopotential in a composite cantilever of piezoelectric dielectrics and nonpiezoelectric semiconductors produced by shear force through e_{15}, *Mater. Res. Express*, 6, 115917, 2019.

[54] G.L. Wang, G.Q. Nie, X.L. Liu and J.S. Yang, Magnetically-induced redistribution of mobile charges in bending of composite beams with piezoelectric semiconductor and piezomagnetic layers, *Arch. Appl. Mech.*, 91, 2949–2956, 2021.

[55] L. Yang, J.K. Du, J. Wang and J.S. Yang, An analysis of piezomagnetic-piezoelectric semiconductor unimorphs in coupled bending and extension under a transvers magnetic field, *Acta Mech. Solida Sin.*, 34, 743–753, 2021.

[56] K. Tsubouchi, K. Sugai and N. Mikoshiba, AlN material constants evaluation and SAW properties on AlN/Al_2O_3 and AlN/Si, in: *Proc. IEEE Ultrasonics Symp.*, 1981, pp. 375–380.

[57] F. Ramirez, P.R. Heyliger and E. Pan, Free vibration response of two-dimensional magneto-electro-elastic laminated plates, *J. Sound Vib.*, 292, 626–644, 2006.

[58] A.W. Warner, M. Onoe and G.A. Couqin, Determination of elastic and piezoelectric constants for crystals in class (3m), *J. Acoust. Soc. Am.*, 42, 1223–1231, 1967.

Index

www.ingramcontent.com/pod-product-compliance
Lightning Source LLC
Chambersburg PA
CBHW050557190326
41458CB00007B/2078